SENSES OF PLACE: SENSES OF TIME

Senses of Place: Senses of Time

Edited by

G.J. ASHWORTH
University of Groningen, The Netherlands

BRIAN GRAHAM
University of Ulster, UK

Routledge
Taylor & Francis Group

LONDON AND NEW YORK

First published 2005 by Ashgate Publishing

Published 2016 by Routledge
2 Park Square, Milton Park, Abingdon, Oxfordshire OX14 4RN
711 Third Avenue, New York, NY 10017, USA

First issued in paperback 2016

Routledge is an imprint of the Taylor & Francis Group, an informa business

British Library Cataloguing in Publication Data
Senses of place : senses of time. - (Heritage, culture and
   identity
   1. Group identity 2. Identity (Psychology) 3. National
   characteristics 4. Cultural awareness 5. Nationalism
   I. Ashworth, G. J. (Gregory John) II. Graham, B. J. (Brian
   J.)
   305

Library of Congress Cataloging-in-Publication Data
Senses of place: senses of time / edited by G.J. Ashworth, Brian Graham.
       p. cm. -- (Heritage, culture and identity)
   Includes bibliographical references and index.
   ISBN 0-7546-4189-9
   1. Group identity. 2. Identity (Psychology) 3. Ethnicity. 4. History--Philosophy. I.
Ashworth, G. J. (Gregory John) II. Graham, Brian. III. Series.

HM753.S48 2005
305.8'001--dc22

                                                                              2004066019

ISBN 13: 978-1-138-24845-8 (pbk)
ISBN 13: 978-0-7546-4189-6 (hbk)

Contents

List of Figures

List of Tables

List of Contributors

Bart van der Aa is working on a Ph.D on world heritage in the Department of Cultural Geography, Faculty of Spatial Science, University of Groningen.

G.J. Ashworth is Professor of Heritage Management and Urban Tourism in the Department of Planning, Faculty of Spatial Science, University of Groningen.

K.I.M. van Dam is currently working at the Arctic Centre, University of Groningen. She is completing a PhD research project on regional identity and sustainable development in Nunavut, Canada.

Brian Graham is Professor of Human Geography at the University of Ulster.

Peter Groote is University Lecturer in Cultural Geography at the Faculty of Spatial Sciences of the University of Groningen.

Tialda Haartsen teaches Human and Cultural Geography at the University of Groningen. Her research interests lie in the field of Rural Geography.

Bettina van Hoven is University Lecturer at the Department of Cultural Geography, Faculty of Spatial Science, University of Groningen.

Paulus P.P. Huigen is Professor of Cultural Geography, Faculty of Spatial Science, University of Groningen.

M.J. Kuipers is working on a PhD on the residential uses of designated monuments and conservation areas in the Department of Planning, University of Groningen.

Amanda McMullan is a Research Associate in the Academy for Irish Cultural Heritages, University of Ulster.

Louise Meijering is working on her doctoral thesis on intentional communities in rural areas at the Department of Cultural Geography, University of Groningen.

Kenneth J.S. Miller is Lecturer in Tourism at the Christelijke Hogeschool Nederland in Leeuwarden.

Bryonie Reid is a PhD student in the Academy for Irish Cultural Heritages, University of Ulster.

Carola Simon is a human geographer and worked at the Faculty of Spatial Science, University of Groningen. Currently she is working as a researcher at Scoop Institute for Social and Cultural Development in the province of Zeeland.

Jonathan Stainer is a Research Associate in the Academy for Irish Cultural Heritages, University of Ulster.

Catherine Switzer is working on a PhD in the Academy for Irish Cultural Heritages, University of Ulster.

Yvonne Whelan is a Lecturer in Cultural Geography, University of Bristol.

INTRODUCTION

Chapter 1

Senses of Place, Senses of Time and Heritage

The Editors

Introduction

The overall aim of this book is to examine the relationships between place and time as these are related though the medium of heritage. In defining the discourses of inclusion and exclusion that constitute identity, people call upon an affinity with places or, at least, with representations of places, which, in turn, are used to legitimate their claim to those places. By definition, such places are imaginary but they still constitute a powerful part of the individual and social practices which people use consciously to transform the material world into cultural and economic realms of meaning and lived experience. Senses of places are therefore the products of the creative imagination of the individual and of society, while identities are not passively received but are ascribed to places by people. While commonplace, such statements need re-stating here for two reasons. First, as occurs with nationalist ideologies, people do often assume that identities are intrinsic qualities of landscapes and cityscapes. Secondly, it is not enough to conclude that places are imagined entities. Rather, if individuals create place identities, then obviously different people, at different times, for different reasons, create different narratives of belonging. Place images are thus user determined, polysemic and unstable through time.

This raises a number of general issues. First, given this intrinsic variability in time, through space, and between social groups, it may seem perverse to attempt to generalize at all about a phenomenon that relates ultimately to a particular individual person, moment and location. The concept of 'collective identity', like the notions of 'collective memory' or 'collective heritage', with which it is strongly related, does not supersede or replace individual identity. It does, however, allow generalization and the location of ideas of belonging within political and social contexts.

Secondly, if it is axiomatic that place images are created, then someone creates them for some purpose. This leads directly to the formulation and execution of policy. Place images do not simply come into existence. Instead, they are created by and through processes of identification which are both internal to the individual or group and external in the sense that they are imposed by outside agency. This leads to questions such as: who is identifying and for what purpose? Place images are not generally explicable in terms of a single simple dominant ideology projected from definable dominant producers to subordinate passive consumers. The peoples, the identities, the images and the purposes are just all too plural to be reduced simplistically in this way.

Thirdly, senses of place must be related to senses of time if only because places are in a continuous state of becoming (Pred, 1984). The key linkage in this process is heritage. At the outset, it is vital to understand that this concept does not engage directly with the study of the past. Instead heritage is concerned with the ways in which very selective material artefacts, mythologies, memories and traditions become resources for the present. The contents, interpretations and representations of the resource are selected according to the demands of the present; an imagined past provides resources for a heritage that is to be bequeathed to an imagined future. It follows too that the meanings and functions of memory and tradition are defined in the present. Further, heritage is more concerned with meanings than material artefacts. It is the former that give value, either cultural or financial, to the latter and explain why they have been selected from the near infinity of the past. In turn, they may later be discarded as the demands of present societies change, or even, as is presently occurring in the former Eastern Europe, when pasts have to be reinvented to reflect new presents. Thus heritage is as much about forgetting as remembering the past.

Heritage, Place and Time

It is not the intention here to follow the concept of 'cultural capital' elaborated by Bourdieu (1977). He posits that a ruling élite, upon assuming power, must capture the 'accumulated cultural productivity of society and also the criteria of taste for the selection and valuation of such products' (Ashworth, 1994, p. 20), if it is to legitimate its exercize of power. Thus it can be argued that dominant ideologies create specific place identities, which reinforce support for particular state structures and related political ideologies. However, clearly, this is too constrained a perspective. As we argue, heritage is capable of being interpreted differently within any one culture at any one time, as well as between cultures and through time. Further, while Bourdieu's thesis implies that evocations of official collective memory underpin the quintessential modernist constructs of nationalism and legitimacy, it is also apparent that heritage takes a variety of official (state-sponsored) and unofficial forms, the latter often being subversive of the former.

Thus heritage is seen here as a much more diverse knowledge in the sense that there are many heritages, the contents and meanings of which change through time and across space. Consequently, we create the heritage that we require and manage it for a range of purposes defined by the needs and demands of our present societies. Perhaps the easiest way of conceptualizing this interpretation of heritage is through the idea of representation. Hall (1997) argues that culture is essentially concerned with the production and exchange of meaning and their real, practical effects. 'It is by our use of things, and what we say, think and feel about them – how we represent them – that we give them a meaning' (Hall, 1997, p. 3). Although he is writing specifically of language as one of the media through which meaning is transmitted, heritage can be regarded as an analogous process. Like language, it is one of the mechanisms by which meaning is produced and reproduced. Hall proposes a 'cultural circuit', which can be extended to include heritage. Meaning is marked out by identity, and is produced and exchanged through social interaction in a variety of media; it is also produced through consumption. These meanings further regulate and organize our conduct and practices by helping set rules, norms and conventions:

> It is us – in society, within human culture – who make things mean, who signify. Meanings, consequently, will always change, from one culture or period to another (Hall, 1997, p. 61).

However the synonymy between language and heritage is not precise because the latter also exists as an economic commodity, which may overlap, conflict with or even deny its cultural role. Heritage is therefore a contested concept and quite inevitably so. Tunbridge and Ashworth's thesis of dissonant heritage (1996) represents the most sustained attempt to conceptualize this facet of heritage and its repercussions. Dissonance is a condition that refers to the discordance or lack of agreement and consistency as to the meaning of heritage. For two sets of reasons, this appears to be intrinsic to the very nature of heritage and should not be regarded as an unforeseen or unfortunate by-product. First, dissonance is implicit in the market segmentation attending heritage as an economic commodity – essentially comprising tangible and intangible place products, which are multi-sold and multi-interpreted by tourist and 'domestic' consumers alike. That landscapes of tourism consumption are simultaneously other people's sacred places is one of the principal causes of heritage contestation on a global scale. Secondly, dissonance arises because of the zero-sum characteristics of heritage, all of which belongs to someone and logically, therefore, not to someone else. The creation of any heritage actively or potentially disinherits or excludes those who do not subscribe to, or are embraced within, the terms of meaning attending that heritage. This quality of heritage is exacerbated because it is often implicated in the same zero-sum definitions of power and territoriality that attend the nation-state and its allegories of exclusive membership. In this sense, dissonance can be regarded as destructive but, paradoxically, it is also a condition of the construction of pluralist, multi-cultural societies based on inclusiveness and variable-sum conceptualizations of power. Whether through indifference, acceptance of difference or, preferably, mutuality (or parity) of esteem, dissonance can be turned round in constructive imaginings of identity that depend on the very lack of consistency embodied in the term.

If heritage is contested along several different axes – the temporal, the spatial, the cultural/economic and the public/private, it also functions at a variety of scales in which the same objects may assume – or be attributed – different meanings (Graham et al., 2000; Graham, 2002). The importance of heritage as a concept is linked directly to that of modernist nationalism and the nation-state and the national scale remains pre-eminent in the definition and management of heritage; United Nations Economic, Social and Cultural Organization (UNESCO) world heritage sites, for example, are nominated by national governments. Nevertheless, even when heritage is defined largely in the national domain, the implementation of policies and their direct management is likely to be conducted at the more local scale of the region or city. Hence heritage is part of the wider debate about the ways in which regions are being seen as the most vital sites within which to convene and capitalize on the flows of knowledge in contemporary globalization. Networking, entrepreneurialism, collaboration, interdependence and a shared vision are all vital prerequisites for regional economic regeneration. Simultaneously, other institutions and agencies are also involved in strategies that 'can serve to circulate and capitalize on existing and other sources of knowledge' (MacLeod, 2000, p. 232), heritage among them. Indeed it may well be a critical factor in that heritage creates representations of places that provide necessary time environments within which more essentially economic processes of wealth generation and marketing can be articulated.

It is a key feature of the post-Fordist capitalist society that knowledge is an input and an output in economic activities. Castells (1996) argues that cultural expressions in what he terms the network society are abstracted from history and geography and become predominantly mediated by electronic communication networks. These latter, which allow labour, firms, regions and nations to produce, circulate and apply knowledge, are fundamental to economic growth and competitiveness. Castells sees a world working in seconds while the 'where' questions – such as environmental sustainability – are in long-term, 'glacial' time. Power, which is diffused in global networks,

> lies in the codes of information and in the images of representation around which societies organize their institutions, and people build better lives, and decode their behaviour. The sites of this power are people's minds (Castells, 1997, p. 359).

Heritage is one fundamental element in the shaping of these power networks and in elaborating this 'identifiable but diffused' concept of power. It is a medium of communication, a means of transmission of ideas and values and a knowledge that includes the material, the intangible and the virtual. Even, arguably, heritage professionals constitute, as Castells would have it, one of the global networks that produce and distribute cultural codes. Yet at the core of these ideas lies the key assertion that the global network has diminished place. Certainly Castells (1998, p. 357) admits to the re-emergence of local and regional government as being better placed to 'adapt to the endless variation of global flows' but this also points to heritage being a knowledge that is rooted in place and region. Its narratives may communicate the local to the global network, for example through the representations of international tourism and marketing imagery, but critically, they are often far more intensely consumed as inner-directed or internalized, localized mnemonic structures. The rise of the network society does not necessarily lead to the demise of place; rather it points to a redefinition of place at the scale of the local and the regional at the expense of the national.

The Uses of Heritage

To reiterate, heritage is that part of the past which we select in the present for contemporary purposes, whether they be economic or cultural (including political and social factors) and choose to bequeath to a future. Both past and future are imaginary realms that cannot be experienced in the present. The worth attributed to these artefacts rests less in their intrinsic merit than in a complex array of contemporary values, demands and even moralities. As such, heritage can be visualized as a resource but simultaneously, several times so. Clearly, it is an economic resource, one exploited everywhere as a primary component of strategies to promote tourism, economic development and rural and urban regeneration. But heritage is also a knowledge, a cultural product and a political resource and thus possesses a crucial socio-political function. Thus heritage is accompanied by a complex and often conflicting array of identifications and potential conflicts, not least when heritage places and objects are involved in issues of legitimization of power structures.

The Economic Uses of Heritage

As Sack (1992) states, heritage places are places of consumption and are arranged and managed to encourage consumption; such consumption can create places but is also place altering. 'Landscapes of consumption ... tend to consume their own contexts', not least because of the often assumed 'homogenising effect on places and cultures' of tourism (Sack, 1992, pp. 158-9). Moreover, preservation and restoration freezes artefacts in time whereas previously they had been constantly changing. Heritage is the most important single resource for international tourism. Tourism producers operate in both the public and private sectors. They may be development agencies charged with regional or urban regeneration and employment creation, or they can be private sector firms concerned entirely with their own profit margins. Whichever, tourism producers impose what may well be relatively unconstrained costs on heritage resources. In turn, the relationship between costs and benefits is very indirect. It may well be that the capital from tourism flows back to heritage resources only indirectly (if at all). It follows, therefore, that heritage tourism planning and management has enthusiastically embraced the idea of sustainable development. If heritage is regarded as a resource, sustainability in this context has four basic conditions. First the rates of use of renewable heritage resources must not exceed their rates of generation: in one sense, all heritage resources are renewable because they can be continuously reinterpreted. Their physical fabric, however, is a finite resource, one factor promoting the immense widening of what might be called the heritage portfolio. Secondly, the rates of use of non-renewable physical heritage resources should not exceed the rate at which sustainable renewable substitutes are developed (for example, the substitution of irreplaceable sites or artefacts with replicas). Thirdly, additional marginal users may displace existing users necessitating prioritization.

Finally, the rates of pollution emission associated with heritage tourism should not exceed the assimilative capacity of the environment (Graham et al., 2000). Heritage management is thus implicated in the belated recognition that the growth in personal mobility in the western world cannot be sustained indefinitely. One outcome is the move towards virtual consumption of place-centred heritage.

The Cultural Uses of Heritage

Heritage is simultaneously knowledge, a cultural product and a political resource. In Livingstone's terms (1992), the nature of such knowledges is always negotiated, set as it is within specific social and intellectual circumstances. Thus key questions include why a particular interpretation of heritage is promoted, whose interests are advanced or retarded, and in what kind of *milieu* was it conceived and communicated? If heritage knowledge is situated in particular social and intellectual circumstances, it is time-specific and thus its meaning(s) can be altered as texts are re-read in changing times, circumstances and constructs of place and scale. Consequently, it is inevitable that such knowledges are also fields of contestation.

As Lowenthal (1985; 1996) has argued, this suggests that the past in general, and its interpretation as history or heritage in particular confers social benefits as well as costs. He notes four traits of the past (which can be taken as synonymous with heritage in this respect) as helping make it beneficial to a people. First, its antiquity conveys the respect and status of antecedence, but more important perhaps, underpins the idea of continuity and its essentially modernist ethos of progressive, evolutionary social development. Secondly, societies create emblematic landscapes – often urban – in which certain artefacts acquire cultural status because they fulfil the need to connect the present to the past in an unbroken trajectory. Thirdly, the past provides a sense of termination in the sense that what happened in it has ended, while, finally, it offers a sequence, allowing us to locate our lives in linear narratives that connect past, present and future.

Although Lowenthal's analysis is couched largely in cultural terms and pays little attention to the past as an economic resource, it is helpful in identifying the cultural – or more specifically – socio-political functions and uses of heritage. Building on these traits which can help make the past beneficial to people, Lowenthal sees it as providing familiarity and guidance, enrichment and escape but also, and more potently, validation or legitimation. This latter trait is particularly associated with identity in which language, religion, ethnicity, nationalism and shared interpretations of the past are used to construct narratives of inclusion and exclusion that define communities and the ways in which they are rendered specific and differentiated (Donald and Rattansi, 1992; Guibernau, 1996). Central to the concept of identity is the Saidian idea of the other, groups – both internal and external to a state – with competing – and often conflicting – beliefs, values and aspirations. These attributes of otherness are fundamental to representations of identity, which are constructed in counter-distinction to them. If yours is the only culture, identity or heritage how do you recognize and demarcate it?

The past validates the present by conveying an idea of timeless values and unbroken narratives that embody what are perceived as timeless values. Thus, for example, there are archetypal national landscapes, both urban and rural, which draw heavily on geographical imagery, memory and myth. Continuously being transformed, these encapsulate distinct home places of 'imagined communities' (Anderson, 1991), comprising people who are bound by cultural – and more explicitly – political networks, all set within a territorial framework that is defined through whichever traditions are currently acceptable, as much as by its geographical boundary. Although many contemporary regions lack this sense of a fixed entity set in history and time, it is apparent that some are evolving as culturally defined bounded spaces in which the 'region-state' aspires to emulate the national state. Both may depend on traditions and narratives that are invented and imposed on space, their legitimacy couched in terms of their relationship to particular representations of the past. In these constructs, the city – particularly the national or regional capital – becomes a landscape that embodies what is defined as official public memory marked by its morphology (the ceremonial axis, the victory arch), monuments, statuary and street names. This urban landscape, in turn, becomes the stage-set for national and regional spectacle, parades and performances.

Implicit within such ideas is the sense of belonging to place that is fundamental to identity. Lowenthal sees the past as being integral both to individual and communal representations of identity and its connotations of providing human existence with meaning, purpose and value. Such is the importance of this process that a people cut off from their past through migration or even by its destruction – deliberate or accidental – in war, often rebuild it, or even 'recreate' what could or should have been there but never actually was. European cities, for instance, contain numerous examples of painstakingly reconstructed buildings that replace earlier urban fabric destroyed in World War II. In the Polish city of Gdansk (formerly Prussian Danzig), for example, the Gothic/Baroque city centre, largely destroyed in World War II, has been reconstructed, not least to link the heritage patrimony of the post-war Polish state to the medieval era before the city became part of the Hanseatic League (Tunbridge, 1998).

Inevitably, therefore, the past as rendered through heritage also promotes the burdens of history, the atrocities, errors and crimes of the past which are called upon to justify the atrocities of the present. Lowenthal (1985) comments that the past can be a burden in the sense that it often involves a dispiriting and negative rejection of the present. Thus the past can constrain the present, one of the persistent themes of the heritage debate being the role of the degenerative representations of nostalgic pastiche, and their intimations of a bucolic and somehow better past that so often characterize the commercial heritage industry with supposed deleterious results to society and economy (see the well known diatribe of Hewison, 1987, on the damage created by the dominance of a backward looking vision).

Themes and Structure of the Book

In pursuing the general issue of the relationships between senses of place and time and heritage, this book draws upon some 14 case studies, which variously discuss examples drawn from Ireland (both north and south), The Netherlands, Canada, Germany and Mexico. The conscious emphasis on Northern Ireland and The Netherlands is a reflection of the ways in which they stand as exemplars of the broader issues concerned with place, time and heritage. In the former, the debates surround unagreed identities and the dissonance that exists between unofficial and unofficial representations and narratives of place. In The Netherlands, conversely, place, time and heritage are dynamically linked in the broader realm of spatial planning, the country providing what is perhaps the best example of the conflicts and tensions that emanate from the integration of senses of place and time into the physical planning process.

The case studies are linked by the three themes that provide the book's structure. These are:

- Creating senses of place from senses of time;
- The public/official creation of place identities;
- Insiders and outsiders.

Within these broad thematic clusters, the individual chapters consider the creation of place identities at local, regional, national and international scales and focus on aspects of the following questions:

- What is a sense of place? How do senses of time interact with senses of

place?
- How do senses of place interact with senses of ethnic/ and cultural identity?
- Who are the actors creating senses of place and what conflicts of interest exist between them?
- At which spatial scales do senses of place exist? How do identities at international, national, regional and local scales interact?
- How and why do all of these variables change through time?

Ultimately, we conclude that heritage is the medium through which senses of place are created from senses of time. However, the processes are contested – and inevitably so – because of their plurality and the continuous interplay of the official/ unofficial and insider/outsider dichotomies that characterize every single manifestation of heritage.

References

Anderson, B. (1991), *Imagined Communities: Reflections on the Origins and Spread of Nationalism*, Verso, London.

Ashworth, G.J. (1994), 'From History to Heritage – From Heritage to History', in G.J. Ashworth and P.J. Larkham (eds), *Building a New Heritage: Tourism, Culture and Identity in the New Europe*, Routledge, London, pp. 13-30.

Bourdieu, P. (1977), *Outline of a Theory of Practice*, Cambridge University Press, Cambridge.

Castells, M. (1996), *The Rise of the Network Society*, Blackwell, Oxford.

Castells, M. (1997), *The Power of Identity*, Blackwell, Oxford.

Castells, M. (1998), *End of Millennium*, Blackwell, Oxford.

Donald, J. and Rattansi, A. (eds) (1992), *'Race', Culture and Difference*, Sage/Open University, London.

Graham, B. (2002), 'Heritage as Knowledge: Capital or Culture', *Urban Studies*, 39, pp. 1003-17.

Graham, B., Ashworth, G.J. and Tunbridge, J.E. (2000), *A Geography of Heritage: Power, Culture and Economy*, Arnold, London.

Guibernau, M. (1996), *Nationalisms: The Nation-state and Nationalism in the Twentieth Century*, Polity Press, Oxford.

Hall, S. (ed.) (1997), *Representation: Cultural Representations and Signifying Practices*, Sage/Open University, London.

Hewison, R. (1987), *The Heritage Industry: Britain in a Climate of Decline*, Methuen, London.

Livingstone, D.N. (1992), *The Geographical Tradition*, Blackwell, Oxford.

Lowenthal, D. (1985), *The Past is a Foreign Country*, Cambridge University Press, Cambridge.

Lowenthal, D (1996), *The Heritage Crusade and the Spoils* of History, Cambridge University Press, Cambridge.

MacLeod, G. (2000), 'The Learning Region in an Age of Austerity: Capitalizing on Knowledge, Entrepreneurialism and Reflexive Capitalism', *Geoforum*, 31, pp. 219-36.

Pred, A. (1984), 'Place as a Historically Contingent Process: Structuration and the Time Geography of Becoming Places', *Annals of Association of American Geographers*, **74**(2), pp. 279-97.

Sack, R.D. (1992), *Place, Modernity and the Consumer's World*, John Hopkins University Press, Baltimore MD.

Tunbridge, J.E. (1998), 'The Question of Heritage in European Cultural Conflict', in B. Graham (ed.), *Modern Europe: Place, Culture, Identity*, Arnold London, pp. 236-60.

Tunbridge, J.E. and Ashworth, G.J. (1996), *Dissonant Heritage: The Past as a Resource in Conflict*, Wiley, Chichester.

THEME I:
CREATING SENSES OF PLACE
FROM SENSES OF TIME

Introduction to Theme One

The Editors

The first block of chapters introduces the central idea of senses of place in relation to senses of time. It raises the relation between place identity and social group identity and between the individual and the communality. The process of identification of people with places and the role of heritage, as a contemporary use of pasts, is the leitmotif of all five chapters.

Huigen and Meijering's [2] initial assertion that identities are ascribed to places and are not an inherent characteristic of them is neither new nor surprising. None the less, it is still of central importance to the themes of this book. The focus of their case study of De Venen where, as in many other such regions in Europe, identity has been challenged by changing economic, social and settlement circumstances. A local identity, based upon an agricultural economy and a rural society, is changing inevitably as that economy and society itself alters. The demise of the conditions that shaped the old identity is clear but the consequences of these changes for a new representation of belonging are not yet apparent.

Simon's analysis of another Dutch region, the Waterland area of North-Holland [3], views the problem of place identity from the position of the place-makers, those who, for one reason or another, wish to establish strong senses of identification between peoples and their localities. This goes further than merely fostering senses of belonging among local residents. It may entail the commodification of places for consumption externally through the creation of marketable place products from previously non-marketed resources. In the case of Waterland, a particular rural identity is being used in the marketing of various agricultural products. If places can becomes products in this way, then they can also be part of a branding process where the product is so strongly associated with the place as to incorporate presumably attractive representations of the region into the product itself. Here, the attempt is made to associate largely agricultural products with an imagined set of regional characteristics that are assumed to convey desirable qualities. Even senses of time are evoked so that, by association, the products and producers acquire attributes of continuity, reliability, and craftsmanship, all of which are incorporated into the product brand. The process can of course be reversed as places may use local products as an element in their local identity, projecting themselves as the homeland of a well-established product. The boundaries between the promotion of commercial enterprises and the place marketing by official bodies become blurred as their goals and methods converge. Similarly the distinction between place identities intended for external consumers and those for internal consumption becomes more difficult to discern as the two markets interact.

The naming of places is both a necessary means of recognition and communication but also a fundamental means of laying claim to territory. The process of naming is more than a value-free description of a point in space, being a means of expressing and fostering senses of place and linking these with selected aspects of the past. Using the example of rural Northern Ireland, Reid [4] examines the relationships between identity and memory through the naming of local places. She acknowledges that naming can be part of broader processes of inclusion and exclusion when linked to particular historical narratives in a divided or unagreed society. While local names may be indicative of diverse cultural influences, they can also be suborned to interpretations that reject pluralist notions of consociation in favour of singular ethnic figuring of space and place. Clearly, this does occur in Northern Ireland where the material marking of placenames in the actual landscape can be part of a broader claim to ethnic territoriality. But in her analysis of the Townlands Campaign in Northern Ireland, Reid shows, too, that the marking of local place remains of such fundamental importance that the process and its associated practices may themselves encourage divided peoples to join together in order to protect and perpetuate their named localities.

If, as argued in the previous chapters, place identity is expressed intentionally or incidentally through aspects of the material built environment, or even through the naming of place, then the obvious question arises as to what happens to these environments when identities alter because of the dynamics of social change through time. Monuments, for example, are endowed with various meanings and their erection or designation is a deliberate act of collective commemoration designed to convey particular messages. The content of these transmissions can be both official – in the sense of promoting statist ideology – or unofficial – in the sense of representing difference or even resistance to the state – depending on the provenance of the monument. While these statements may be self-evident, the reaction of monuments to social change is less predictable. There are a number of possibilities. Existing monuments may be removed and replaced; they may be re-designated and their meanings reinterpreted to express new meanings; or they may simply become ignored and rendered all but invisible, their meanings lost through being irrelevant or unreadable. Whelan [5] examines these more general issues in the context of the city of Dublin which, in little over a century, has been projected through its monumental landscape as an Imperial showcase, the provincial capital of a constituent British kingdom, the capital of a Catholic Gaelic nationalist republic and, more recently, a multicultural European city. She demonstrates how particular aspects of Dublin's cultural landscape, chief among them public statues, street names, architecture and urban design initiatives, have served as significant sources for unravelling the geographies of political and cultural identity in Ireland. While once again these processes have resonances of the issues raised by Reid in Chapter 4 concerning the naming and claiming of place, significantly the most recent monument to be erected in the city, the 'Spire of Dublin', is almost definitively postmodernist in its utter simplicity, ahistorical nature and complete lack of any political association, all of which allow a multiplicity of readings of its symbolic place. Its only meaning may be the absence of meaning.

The final chapter in Part One is somewhat different, being the description and application of a technique of spatial analysis. McMullan [6] confronts and attempts to solve the problem that much of the substance of the other chapters in this book involves disparate forms of qualitative data, which are often considered unsuited to quantitative analysis. While the advent of Geographical Information Systems (GIS) permits far more precise quantitative analysis of place, most cultural and many historical geographers remain resistant to the methodology. This negates the ways in which GIS can be used to integrate the qualitative and the qualitative in a finely differentiated spatial mesh, which seems readily attuned to the increasing emphasis and focus on, and internalization of, the local. Heritage as we have defined it, is of course incapable of being mapped but the resources, actual and potential, from which it is composed, can be investigated through cartographical analysis. McMullan demonstrates the ways in which GIS can be used to enhance data collection, description and analysis in an ethnological research project concerned with folk practices in Ireland. She demonstrates how GIS can overcome the interpretative barriers involved in the interpretation of paper-based ethnological maps and argues that resistance to the methodology is itself a function of academic perceptions rather than a comment on the value of the methodology for research in cultural geography and heritage studies.

Chapter 2

Making Places: A Story of De Venen

Paulus P.P. Huigen and Louise Meijering

Introduction

This study attempts to trace the evolution of the identity of De Venen. For this purpose, the 'identity' of De Venen is defined as the ideas that exist about the area known by that name. This area is located in the Green Heart of the Randstad conurbation in The Netherlands (see Figure 2.1). We will outline existing perceptions of De Venen that define not only what the area *is*, but also what it should be. Furthermore, we will describe how the identity of De Venen will evolve in the future. The aim of this study is to indicate the path that the place identity of De Venen may follow over time. Place identity is a central concept in this chapter. We will therefore first set out our vision of the concept by exploring six characteristics of place identity. These will then be applied and discussed in relation to the development of the De Venen area.

In the 1990s, a remarkable change took place in academic and policy debates on rural areas in The Netherlands. Until the 1990s the idea prevailed that, in economic and cultural respect, rural areas in the north and east of The Netherlands were lagging behind the urban agglomerations (especially the Randstad conurbation in the western Netherlands). There was deprivation in these areas and the idea was that this situation could be resolved by involving them in the development of the national agglomerations. To achieve this, part of the development activities had to be transferred to the more 'peripheral' areas. This 'overspill strategy', introduced in the First Policy Document on Spatial Planning in the Netherlands (Ministerie van Volkshuisvesting en Bouwnijverheid, 1960), was developed into plans designed to ensure that rural areas were incorporated into national and international networks. This included establishing infrastructural links to the national agglomerations. The vision underlying the 'overspill strategy' can be described as a traditional approach to area development.

Thinking on area development changed in the sense that a new vision, the 'identity strategy', evolved in addition to traditional development ideas. The basic idea was that the qualities of rural areas and their identities should be taken as the starting point, rather than the situation in the national agglomerations. Area development would be realized by strengthening and profiling those qualities and identities, instead of copying developments in the national agglomerations (European Conference on Rural Development, 1996; Provincie Groningen, 2000; Raad voor het Landelijk Gebied, 1999a; 1999b). The call for greater recognition of local characteristics and identity in area development was not only related to the 'peripheral' areas (that is, areas outside the urban agglomerations) but also rural areas within the Randstad conurbation under pressure from spatial demands. These areas were pictured as 'green' and undeveloped areas. Cities, too, experience a growing need to emphasize their identity in the face of overwhelming 'McDonaldization' and the accompanying trend towards uniformity. As this example illustrates, the increasing emphasis on regional identity can be partly explained as a reaction to the process of globalization (Giddens, 1998; Harvey, 1989). Globalization, which can be defined as the interrelationships between different places that evolve through common processes of economic, political, cultural and environmental change, has led to increased mobility of people, goods and information. Furthermore, we increasingly feel that we live in a progressively more unstable and uncertain world. This feeling of insecurity is being fed by the increasing amount of information about other, strange places and people and societies that becomes available in the globalization process. This perception in itself has led to a growing desire for stable, certain places with a coherent identity (Massey, 1995). People are searching for a point of reference in their own environment, which regional, local or place identity can provide (Van der Borgt et al., 1996; Massey, 1995).

It is not always clear what exactly is meant by 'place identity'. Sometimes it is described in terms of culture, particularly local music, folklore (for example, traditional costumes) or culinary specialities. Often, a place derives its identity from the landscape or built environment. Place identity is also derived from regional economic activities (horticulture in the 'greenhouse city', port activities in Rotterdam, dairy farming in the Green Heart) or products (cheese from Leerdam, Gouda and Edam). That identity is used for 'place marketing', in tourism and attracting new residents or investors. A place has to have a 'face' to succeed in marketing terms. Heritage may be the outstanding place characteristic, which is related to the 'face' of a place. This summary of the various approaches to and uses of the term 'place identity' shows that it is not a completely transparent concept. Therefore, we shall first explore this concept.

The Concept of Place Identity

In this study, the term 'place' is a synonym for a region or area. In all cases it denotes spatial entities above the local level of varying size. A region or area can contain several places. Although policymakers use the term *place identity* in different, often implicit senses and contexts, the academic world appears to have reached a broad consensus on six aspects of place identity (Groote et al., 2000).

First, place identity is a social construct. It is not 'something out there, waiting to be discovered'. It is not an objective or natural 'given', but something that is attributed to a place by people. We refer to the process of attributing an identity to a place as the construction of place identity. The result of this process is a place identity that is based on a point in time and the social construction of the place, in other words: its socio-cultural significance. Giving a name to a place is one of the basic activities in this construction process (Paasi, 1991; Simon et al., 2001). The name distinguishes the place from other spatial entities. A place is created when we attribute to it the quality of being 'distinct' from another location and give it a name. If that name has an additional meaning that refers to another feature of the place, the place identity is reinforced. Places lose their identity when they become an indistinguishable part of a larger whole. After such an integration process, a place is no longer seen as separate or distinct. Rose (1995) distinguishes two processes of identity attribution. First, identifying *with* a place involves attributing an identity to it that is based on positive feelings of connection with that place. The attribution of an identity implies that being in that particular place makes one feel 'at home' and 'at ease'. Identifying with a place is partly a positive thing because the person also derives part of his own personal identity from it. Secondly, identifying *against* involves attributing an identity to a place that is based on negative feelings. Anti-urbanism is an example of this because it attaches qualities such as 'dangerous', 'overcrowded' and 'polluted' to urban locations. Typically, 'identifying against' often relates to 'other' – less familiar – places, which tells more about the perceptions and culture of the person than about the place. The attribution process may take time. This means that an identity evolves continually and consists of several chronological 'layers'. It is a social construct that embodies elements from the past and the present in the desired situation. Attributing an identity to a place has two main implications. Initially, when attributing an identity, a person attempts to express the 'true' nature of a place, or those features that define its essence for her or him. This identity is established through the way in which a person communicates about a place. Conflicts surrounding the identity of a place mean that its essence is denied. It is no longer the 'same' place. Conflicts contribute to the conservation of a place identity. The stronger place identity becomes, the more difficult it is to change it and the greater the number of conflicts. Subsequently, when an identity is attributed to a place, that place becomes partly enclosed; other place identities are excluded. Establishing a place identity is therefore partly a process of exclusion, of shutting out.

The second aspect of place identity is that it is based on the characteristics of the place. Place-identity attribution is not a random process. Actors (people in their organizational setting: households, businesses, institutions) attribute an identity in order to achieve certain implicitly or explicitly stated goals. The chosen identity depends on the goals of the actor in question. Actors link the attainment of their goals to the perceived features of the place through place identity. In other words, the actor bases the place identity on the distinguishing features of the place, which are often considered as characteristics and identity markers. The characteristics of the spatial environment/surroundings can also be seen as features of the place.

Thirdly, place identity is to a large extent based on the past. Since the future is unknown, the past plays an important role in identity attribution. Place identity is largely based on the characteristics of the place, as it was perceived in the past. When judging whether a new development fits with a place identity, people use arguments such as '... has always been here' or 'that would fit in perfectly', or 'that doesn't fit in at all' with what is customary in that place. The history of a place can be seen as a lucky dip. Everyone reaches into it for 'facts' to support their own particular goals. For every historical 'fact' used in support of the argument that a development is in line with tradition, another historical 'fact' can be found to prove exactly the opposite. The process of identity attribution is often influenced by nostalgic sentiments. An identity is based on perceptions not only of what a place *is*, but what it *should be*. This is particularly evident in the identities of urban places and regions, where the traditional rural way of life has been lost (Jensma, 1997).

In the fourth place, place identity is debatable. Every society is composed of different actors with different goals, who therefore attribute different identities to a place. This means that place identity is always open to debate. Actors who are powerful in terms of authority and/or resources can impose their dominant place identities at the expense of other actors. Different actors will always attribute a different identity to a place. It is therefore advisable to refer to place identities in the plural. It should be noted here that the debatable nature of place identities does not always result in a 'struggle'. Here, 'debatable' refers to a certain degree of divergence in an apparently undisputed place identity.

The fifth feature of place identity is that it is attributed within, and characterized by, a particular context. Here we can identify at least two contexts. The first is the spatial context or location, which can also be labelled with the classic socio-geographical term 'situation'. The second context is the socio-cultural context, which consists of the prevailing norms and values of society. This includes social and economic circumstances and relationships, as well as spatial planning trends relating to the structure and function of places.

The final aspect of place identity is that its identity attribution is a continuing process in which new actors establish themselves and goals and ideas change. The context within which the actors are located also changes. Place identities evolve continually, but at the same time established identities are confirmed. They are then reproduced in the conduct of the place and the way in which it is represented through communication information exchange. Identities are constructed and reconstructed. This process follows a certain path over time. Along that path we find identity markers, characteristic objects and events that define the evolution of that place identity. The future construction of place identities is determined by two factors. The first factor is evolutionary. Every situation in the present and in the future is inevitably rooted in the past. A current place identity and its historic component form the basis for the future identity, which in turn will embody many aspects from the past. The second factor is unexpected events, the substance of which, by their very nature, cannot be known. Although the evolutionary path is a determining factor for the future of a place, it should be balanced against the occurrence of unexpected developments.

De Venen

The six aspects of place identity discussed above make place identity very suitable for tracing the evolution of a place. This study will continue by examining the evolution of De Venen on the basis of the area's place identity. Recently, the story of De Venen has been beautifully told in word and image in Buissink and Fey (2001) who point out that:

> The number of publications on the Green Heart in recent decades is staggering. Almost every institution and organization that has anything to do with the environment, urban planning, transport and traffic, tourism or agriculture has made itself heard. For years, specialists and stakeholders

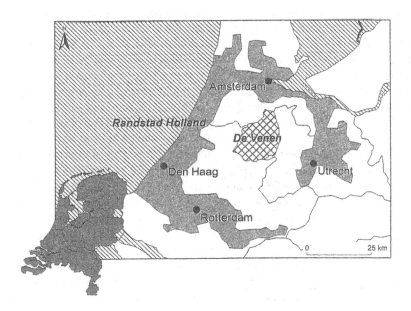

Figure 2.1 Location of De Venen in the Randstad conurbation

have set out their future vision for the area in letters and columns in the newspapers (Buissink and Fey, 2001, p. 67).

This statement indicates that the story of De Venen is a multiple one and that that the future of it is contested. We shall examine De Venen from the perspective of the six aspects of place identity identified above.

First, the dominant socially constructed identity attributed to De Venen is that of a peat grassland area located within the Randstad conurbation. This identity consists of two parts: a landscape characteristic 'peat grassland area' and spatial context characteristic 'in the Randstad'. The landscape characteristic, 'peat grassland', can be described as follows: peat grassland areas are open, wet meadow landscapes, often with elongated parcelling patterns and high water levels. Peat grasslands are important for agriculture, nature, the landscape, recreation and tourism (Ministerie van Landbouw, Natuurbeheer en Visserij, 1992, p. 19). The history of peat extraction and agriculture are determining factors for the place characteristic 'peat grassland' (Berendsen, 1997). The history of peat extraction is clearly expressed in the identity attributed by Buissink and Fey (2001, p. 1): 'The hand of man [sic] is visible everywhere in this cultural landscape. This makes it unique in the world'. With regard to agriculture, it is primarily dairy farming that shapes the landscape of green meadows and ribbon development with farms (De Pater et al., 1989): 'for most people, the area is 'farmland decorated with patches and strips of nature' (Buissink and Fey, 2001, p. 68).

The spatial context characteristic, 'in the Randstad', figures De Venen as part of the Green Heart of the Netherlands, an area where land use is under heavy pressure from urbanization and economic growth (De Pater et al., 1989; Van der Ploeg, 2001). The relatively high land prices are an indication of this. The economic pressure to urbanize relates to the functions of business, living and recreation. The heavy pressure from the spatial planning sector results from landscape protection, environmental rehabilitation (that is, establishing the national ecological infrastructure) and preserving green open spaces in the Randstad conurbation. In such over-pressurized areas, displacement processes occur whereby actors and functions that are strong in terms of economics or policy displace their weaker competitors, and related forms of land use. Secondly the place identity of De Venen is largely based on landscape and spatial context. The landscape is characterized by different parcelling areas, uplands, small river levees, long narrow dredging ponds, drained lakes and the water management systems (Borger et al., 1997; Burger and Jungerius, 2001). The spatial context of the area plays an important – perhaps even dominant – role in the identity attribution. Two elements are involved. The first is the relatively open character of the area and associations of tranquillity, space and greenery (Visscher, 1975) in contrast to the surrounding built-up Randstad, which is perceived as busy, exciting and a source of culture. The second important element is the proximity to the infrastructure of the Randstad (De Pater et al., 1989). This is evident from the many recreational facilities and high level of 'import', resulting in many expensive houses and boats ('luxury'). Water in De Venen underpins the importance of landscape and location as anchors of place identity. The presence of water reinforces the landscape characteristics 'not built up ' and 'open space', and also supports recreation activities. The effect of this is to reinforce the aspects of the place identity that depend on the spatial context.

As can be derived from the third aspect of place identity, the past is an important factor in the place identity of De Venen:

> Right up to the present day, the character and form of the Randstad and the Green Heart have been shaped by the sand, the clay and the peat bogs drained by human labour. History can be read from the landscape (Buissink and Fey, 2001, p. 16).

The past, as manifested in the landscape, is used to bring sturdiness in the identity construction of De Venen. The past is also used to underpin essential activities of the place: 'dairy farming has traditionally supported and managed the features of the landscape in the peat grassland area' (Ministerie van Landbouw, Natuurbeheer en Visserij, 1992, p. 19). The area is characterized as a reclaimed peat area. The land was reclaimed from the relatively high dikes bordering the rivers, where the farms were built as well. This resulted in long, narrow pieces of land and elongated village structures. As in the reclaimed lake areas, livestock farming quickly developed as the main form of agriculture because groundwater levels were high and drainage was difficult, particularly in the low areas (Berendsen, 1997; De Bruijne, 1939; De Pater, 1989; Visscher, 1975). The history of peat extraction and agriculture – primarily livestock farming – became an essential element in the place identity of De Venen. The area is associated with long rectangular strips of land and grazing cows. These landscape features dominate the appearance of De Venen. Van der Ploeg (2001) has pointed to a number of constants in the history of the peat grasslands, and in their agricultural history in particular: integration into the market economy, farmers struggling with difficult conditions and commercial practices imposed by nearby urban areas as producing for the urban market, selling the products in the cities). In the past, too, the spatial context of De Venen played an important role.

Fourthly, the place identity of De Venen is very much contested. Two interrelated debates about the future identity of De Venen can be discerned. Initially there is a conflict between agriculture and recreation/nature identities. The supporters of recreation and nature want to see an increasingly prescriptive role, while many farmers do not want to give up their businesses. Again, anti-urbanization policy conflicts with the increasing demand for housing. The government places strict limits on housing development, although the demand for houses exceeds the supply. A process of 'hidden' urbanization can be observed in the form of development in existing built up ribbon developments. This sometimes involves the conversion of old farms, or the construction of a new house for the parents of a farmer who has taken over the farm.

In his analysis of planning in the Green Heart, Lingbeek (1998) indicates that various actors represent various interests. He identifies three interests, Green-Randstad, local-authority and regional. The Green-Randstad interest involves the idea that the Green Heart is a green development area for the Randstad conurbation (Lingbeek, 1998, p. 71). 'Tranquillity, greenery, space, nature, and clean water are the key factors' (Stuurgroep de Venen in Lingbeek, 1998, p. 107). A restrictive policy will have to be implemented for expansion in the Green Heart; it must be developed into an area for nature and recreation. Local-authority interests are working towards a 'green urbanization', which achieves a cohesive balance between agricultural, nature-related and tourist/recreational activities. Housing development contributes towards an optimal result if it enhances the natural landscape of the area or increases the feasibility of developing landscape and nature (Projectbureau De Ronde Venen, 1995) Regional interests include 'agricultural nature management' in which farmers can enhance the qualities of the landscape and environment through sustainable agricultural practices.

Each of the three interest groups has attributed its own place identity to De Venen. The Green-Randstad group ascribes a traditional place identity to 'a green place' in which De Venen is seen as part of the Green Heart and as a counterbalance to the Randstad cities. The local authorities group ascribes an identity of 'green urbanization' including a desirable residential area where people can pursue a quasi-rural lifestyle (Van Dam, 2002). The regional group favours an identity of 'green land-use' in which agriculture functions as the 'caretaker of green spaces' and the provider of the desired environment.

Fifthly, place identities of De Venen are contextualized. The first relevant context is the physical/spatial context, that is *where* De Venen is situated. The location of De Venen in relation to national and regional concentrations of socio-spatial activities is essential for its place identity. The relative proximity of Schiphol Airport, and the major cities of Amsterdam and Utrecht influence developments in the housing market, the labour market, the use of facilities and opportunities for recreation in De Venen. Of the three partially conflicting place identities, particularly 'green undeveloped place' (open space) and 'green urbanization' (space for housing) characterize the location of De Venen. The second context is the socio-cultural context, which can be separated into two components. One component relates to the socio-cultural (and also economic) circumstances in society as a whole. These include the economic situation, the nature of housing demand, the amount of leisure time, the level of interest in nature and the landscape. When the economy is prospering, more people are able to fulfil their individual housing requirements. There is stronger demand for larger homes and for homes in the countryside, and this increases the pressure on areas such as De Venen. The second component relates to trends in spatial planning and policy. Which type of city is currently in favour with planners: the compact city, the network city, the urban hub, the ribbon city or the sustainable city? Inevitably, planning trends and socio-cultural circumstances in society as a whole will interact to some extent. Buissink and Fey (2001, p. 29) describe the influence of the socio-spatial context as follows:

> In the course of the twentieth century, policy on nature and the landscape tended to change fairly often. Sometimes the principle prevailed that green space should make way for spacious homes and urban parks for the Randstad population, and at other times the theory prevailed that further encroachment on the landscape was unacceptable for the purpose of providing recreational space for city-dwellers.

The identity attributed to De Venen as a place for green land-use also shows some influence of the prevailing socio-cultural context. Growing pressure from social and agricultural developments means that farmers are increasingly confronted with the socio-cultural and physical context. Agriculture is becoming ruralized again (Huigen and Strijker, 1998): after the Second World War, agriculture experienced a period of economic expansion whereby commercial operations became disconnected from the physical and social rural environment. Villagers no longer came to work or buy products at the farm. The importance of the physical environment was related to its potential for cost-effective farming. 'Agribusiness' grew away from the rural context. This phenomenon is referred to as the de-ruralization of agriculture. Currently, a process of ruralization is taking place and the relationship between agricultural commerce and the rural context is being re-emphasized. This illustrates that place identity is a process, its sixth aspect. The physical environment is important in terms of nature conservation and recreation. In addition, the relationship between commercial operations and the community is becoming stronger through the sale of quality agrarian produce. Promotional and information campaigns (such as Open Days) are helping to strengthen the links between farming and rural communities. The process of ruralization in agriculture is leading to a closer relationship between agricultural commerce and its socio-economic and physical environment. Van der Ploeg (2001, p. 191) refers to this phenomenon as agrarian diversification.

Discussion

Place identities are constructed and reconstructed. The question is: which place identities are likely to dominate the future of De Venen? We have indicated that three place identities apply to the area: green undeveloped; green urbanization; and green land-use. In our view, there are indications that the most obvious place identity for De Venen in the future is as an area of 'green urbanization'. This is based on the following arguments. Economic pressure to urbanize is considerable. Protecting any sizeable area of nature and landscape against such pressure is an immensely difficult task, except for 'small islands', 'flower tubs', and 'nature gardens'. According to Buissink and Fey (2001, p. 28):

> nature conservation has always drawn the shortest straw. The desire to protect was hardly present during economic recessions. When such a need arose during economically prosperous periods, the divided "green front" was too weak to stand up to commercial interests.

Here, for the sake of accuracy, we should mention that Buissink and Fey (2001, p. 28) go on to state that 'recently, the social consensus on the need to preserve the landscape appears to be increasing'.

A restrictive policy in areas that are under pressure to urbanize leads to displacement, whereby economically strong actors push out their weaker counterparts. These are enclosed settlements, where growth or expansion is blocked. Driven by cultural and ecological conservatism, this policy is geared towards preserving open spaces and preventing fragmentation. According to Driessen et al. (1995, p. 38), the landscape is becoming a green residential environment, a park for the wealthy. In a capitalist market economy, displacement is apparently inevitable. The question is: to what extent will it occur? Agriculture in De Venen is under economic pressure due to the unique production conditions of the peatland area and the relatively high land prices in the Randstad, both of which exert upward pressure on cost prices (Van der Ploeg, 2001). Economic perspectives exist in the form of agricultural diversification, particularly in the recreation sector and service provision for city-dwellers. In many cases this will mean further development and urbanization. However, we believe that the economic potential in nature and landscape conservation will prove to be largely marginal and vulnerable. This vulnerability is due to the fact that evaluation of nature and the landscape is anchored in an economic and social context.

It is our view that the 'hidden' urbanization, which is already evident, will intensify. The building density in ribbon developments will increase and there will be greater urbanization for the purpose of supporting the recreation function. Cycle paths through the outlying areas, landing stages for canoes, asphalt car parks for sports facilities in the polders, passing-places for oncoming traffic, separate cycle paths for safety reasons: De Venen will be a busy place. Perhaps the most important reason for assuming that the dominant identity of De Venen will be as a place for 'green urbanization' is that the place identity 'Green Heart', with De Venen as an integral part of it, is referred to in policy. The following quote illustrates this:

> The term "Delta" expresses actual topographical, ecological and landscape qualities, while the term "metropolis" expresses an expectation of innovation and ambition. The new name "Delta metropolis" is therefore a welcome alternative to the old Randstad conurbation/Green Heart dichotomy (Ministerie van Volkshuisvesting, Ruimtelijke Ordening en Milieubeheer, 2002, p. 22).

This quotation makes it clear that the Green Heart and the Randstad conurbation are seen as part of a larger Delta metropolis, which replaces the Randstad–Green Heart dichotomy. Green urbanization within the Delta Metropolis would appear to be the most likely route to the future for De Venen.

Places disappear and lose their identity when they are no longer distinguished from their context, when they vanish in their context. This may happen when places become part of a larger whole. The Green Heart is becoming a series of parks with a number of green residential neighbourhoods within the Delta metropolis. The people who live in these areas can enjoy a pleasant 'green' lifestyle. One of those neighbourhoods is called De Venen.

Acknowledgements

The authors would like to thank Tamara Kaspers-Westra for her cartographic work.

References

Berendsen, H.J.A. (1997), *Landschappelijk Nederland*, Van Gorcum, Assen.

Borger, G., Haartsen, A. and Vesters, P. (1997), *Het Groene Hart: een Hollands Cultuurlandschap*, Matrijs, Utrecht.

Borgt, C. van der, Hermans, A. and Jacobs, H. (1996), 'Een Groeiende Belangstelling voor de Eigen Regio', in C. Borgt, A. van der Hermans and H. Jacobs (eds), *Constructie van het Eigene: Culturele Vormen van Regionale Identiteit in Nederland*, P.J. Meertensinstituut, Amsterdam.

Bruijne, F.H. (1939), *De Ronde Venen*, Libertas Drukkerijen, Rotterdam.

Burger, J.E. and Jungerius, P. (2001), *Landschap lezen. Wandelen in het Groene Hart: De Ronde Venen*, Uitgeverij Op Lemen Voeten, Amsterdam.

Buissink, F. and Fey, T. (2001), *Dwars door het Groene Hart: Landschapsverkenningen door Tijd en Ruimte*, Uniepers, Abcoude.

Dam, F. van (2002), 'De Kwaliteit van het Wonen op het Platteland', *Tijdschrift voor de Volkshuisvesting*, **4**, pp. 36-41.

Driessen, P.P.J., Glasbergen, P. , Huigen, P.P.P. and Hijmans van den Bergh, F. (1995), *Vernieuwing van het Landelijk Gebied. Een Verkenning van Strategieën voor een Gebiedsgerichte Aanpak*, VUGA Uitgeverij B.V, The Hague.

European Conference on Rural Development (1996), *Rural Europe: Future Perspectives: The Cork Declaration: A Living Countryside*, Cork.

Giddens, A. (1998), *The Third Way: Renewal of Social Democracy*, Polity Press, Cambridge.

Groote, P., Huigen, P.P.P. and Haartsen, T. (2000), 'Claiming Rural Identities', in T. Haartsen, P. Groote and P.P.P. Huigen (eds), *Claiming Rural Identities: Dynamics, Contexts, Policies*, Van Gorcum, Assen.

Haartsen, T., Groote, P. and Huigen, P.P.P. (eds) (2000), *Claiming Rural Identities. Dynamics, Contexts, Policies*, Van Gorcum, Assen.

Harvey, D. (1989), *The Condition of Postmodernity*, Basil Blackwell, Oxford.

30   *Senses of Place: Senses of Time*

Huigen, P.P.P. and Strijker, D. (1998), *De Relatie Tussen Landbouw en Samenleving: een Proces van Afstoten en Aantrekken,* The Hague, Nationale Raad voor Landbouwkundig Onderzoek. NRLO-rapport nr. 97/39, p. 43.

Jensma, G (1997), *Hoe Geert Mak in Jorwerd kwam,* De Gids, June.

Keuning, H.J. (1955), *Mozaïek der Functies. Proeve van een Regionale Landbeschrijving van Nederland op Historisch- en Economisch-geografische Grondslag,* reprint 1998, Regio Project Uitgevers, Groningen.

Lingbeek, C.O. (1998), *De Macht van de Metafoor: een Analyse van de Planning voor het Groene Hart* Van Gorcum, Assen.

Massey, D. (1995), 'The Conceptualization of Place', in D. Massey, and P. Jess (eds), *A Place in the World? Places, Cultures and Globalization,* Oxford University Press, Oxford.

Ministerie van Landbouw, Natuurbeheer en Visserij (1992), *Het Landelijk Gebied de Moeite Waard. Structuurschema Groene Ruimte. Ontwerp-planologische Kernbeslissing,* Ministerie van Landbouw, Natuurbeheer en Visserij, The Hague.

Ministerie van Volkshuisvesting en Bouwnijverheid (1960), *Nota Inzake de Ruimtelijke Ordening in Nederland,* Staatsdrukkerij- en Uitgeverijbedrijf, The Hague.

Ministerie van Volkshuisvesting, Ruimtelijke Ordening en Milieubeheer (2002), *Ambities voor de Deltametropool. Eindrapportage Interdepartementaal Project Deltametropool,* Ministerie van Volkshuisvesting, Ruimtelijke Ordening en Milieubeheer, The Hague.

Paasi, A. (1991), 'Deconstructing Regions: Notes on the Scales of Spatial Life', *Environment and Planning A,* **23,** pp. 239-56.

Pater, B.C. de, Hoekveld, G.A. and Ginkel, J.A. van (eds) (1989), *Nederland in Delen. Een regionale geografie. Deel1: Nederland als Geheel, West- en Zuidwest-Nederland,* De Haan, Houten.

Ploeg, B. van der (2001), *Het Wei-gevoel in het Groene Hart van de Randstad: een Studie Onder Melkveehouders in het Westelijk Veenweidegebied naar hun Bereidheid en moGelijkheden zich te Ontwikkelen van Productieboer tot Plattelandsondernemer,* Ponsen en Looijen, Wageningen.

Projectbureau De Ronde Venen (1995), *De Ronde Venen. Ontwikkelingsvisie. Concept,* Projectbureau, De Ronde Venen.

Provincie Groningen (2000), *Provinciaal Omgevingsplan. Koersen op Kkarakter.* Ontwerp, Groningen.

Raad voor het Landelijk Gebied (1999a), *Made in Holland: Advies over Landelijke Gebieden, Verscheidenheid en Identiteit,* Amersfoort.

Raad voor het Landelijk Gebied (1999b), *Geleid door Kwaliteit. Interim-advies over Landelijke Gebieden en de 5e Nota Ruimtelijke Ordening,* Amersfoort.

Rose, G (1995), 'Place and Identity: a Sense of Place', in D. Massey and P. Jess (eds), *A Place in the World? Places, Cultures and Globalization,* Oxford University Press, Oxford.

Simon, C., Groote, P. and Huigen, P.P.P. (2001), 'Verstreking of Ontstreking?' *Rooilijn,* **24**(1), pp. 16-22.

Visscher, H.A. (1975), *De Nederlandse Landschappen. Ontstaan, Wetenschappelijke Betekenis, Belevingswaarde. Deel 2,* Het Spectrum, Utrecht and Antwerpen.

Chapter 3

Commodification of Regional Identities: The 'Selling' of Waterland

Carola Simon

Introduction: Construction of Regional Identities[1]

The concept of regional identity has become a popular theme in the academic discourse in the last decades. Several authors in the discipline of cultural geography (for example Murphy, 1991; Hoekveld, 1993; Cloke et al., 1998; Graham, 1998; Groote et al., 2000; Knox and Marston, 2001) define regional identities as the specific meanings (including feelings and images) that are attached to a region by an actor or different groups of actors who experience the region in different ways. Regional identity is seen as social constructions, which means that we, as individuals or groups, ascribe identities to it. Allen et al. (1996, p. 2) describe this as follows: 'they are not "out there" waiting to be discovered; they are our (and others) constructions'. Following this constructionist approach (see Hall, 1997), no distinction is made between, as Barke and Harrop (1994) and Pellenbarg (1991) suggest, identity ('what the region is actually like') and image ('the picture of a region shown to the outside'). Identity is not seen as an objective thing, but reconstructed by people and subject to change. This means that identity is a dynamic concept constructed by people. Moreover the line of thought is followed that different actors interpret representations of regions in different ways. Therefore, different identities might be allocated to the same region at the same point in time, and for that reason, we should not speak of 'the' identity of a region, but of identities in the plural. Also actors base the regional identities that they proclaim on perceived characteristics or qualities of a region. As the future is not yet known these characteristics are closely linked to the past. Accordingly, it is 'our' sense of time that in the long run creates senses of place (Jackson, 1995).

In particular actors in the professional discourse – those involved in a professional way with the development of a region (Jones, 1995) – might have ulterior motives with the construction of regional identities. Such actors, for example nature conservancy councils or tourist agencies, use strategic means to reach their goals (Brouwer, 1999). The concept of regional identity might be such a strategic expedient. Actors do construct identities about a region with specific motives, for example to conserve heritage (both natural and cultural-historical), to promote a region to attract more visitors or simply to catch subsidies for regional projects.

The principal aim of this chapter is to examine the application of regional identities by actors in the professional discourse. The perspective is that aspects of regional identities are used in the selling of products. The discussion considers how actors apply regional identities in the selling of products, what their motives are for this usage, and which symbols, activities and products are developed herein. Because actors use identities of regions in, for example, advertisements or material used for tourist promotion, these identities may begin to lead a life of their own, and might be adopted by others and reproduced. As a result, regional identities may be used as commodities, suggesting that they are used in products suitable for consumption.

The Commodification of Regional Identities

Commodities are described as 'objects that are produced for the purpose of being exchanged' (Johnston et al., 2000, p. 95), and the process of commodification can be defined as the production and trading in forms of new, intangible or unmanageable products of commodities. According to Ray (1998) and Kneafsey (2000; 2001), the commodification of, in their case, the countryside involves the upgrading of 'place' through the usage of cultural identities within the process of 'production'. This culture economy approach of rural development refers to 'the territory, its cultural system and the network of actors that construct a set of resources to be employed in the pursuit of the interests of the territory' (Ray, 1998, p. 4). Following these thoughts, the commodification of regional identities deals with the relation between product and place. It concerns the use of regional identities in projects and activities that are associated with the region, and that are turned into 'products to be bought and sold' (Holloway and Hubbard, 2001, p. 154). Products are developed for particular purposes and the identities of a region are means to reach these goals. In this way, identities of a region are used to add value to products. This means that in the process of commodification, identities are not particularly constructed for representing a region but for selling products.

Actors in the professional discourse try to increase the demand for their products by using various cultural and regional symbols, like elements of the landscape, regional cuisine, regional languages and regional art (Ray, 1998). The products to be sold are, for example, concerned with food, specific activities related to the region, or the region itself when sold as a tourist area. The products might exist already or have existed, and are in this way related with the past. However, it is also possible that the products are newly developed whereby the connection with a region is recently created or invented. Here, the assumption is that different actors deploy different identities in the selling of their products. For instance, a recreation manager sells coffee and overnight stays while a nature conservator sells the beauty of the landscape. Actors use those identities that are most helpful in selling their products.

Several authors argue that the commodification of regions in marketing strategies continues to flourish (Bell and Valentine, 1997; Mitchell, 1998). Examples are the occurrence of elements of a regions culture, landscape and history in regional agricultural products or material used for tourist promotion. The development of regional quality products is, in this perspective, popular in various European countries, like the Netherlands (Vlieger et al., 1999), the United Kingdom (Ilbery and Kneafsey, 2000), France (Bessière, 1998) or Ireland (Markwick, 2001). Similarly, various tourism commodities, such as souvenirs, are produced and consumed as 'authentic' regional experiences (Halewood and Hannam, 2001). In this chapter, I focus on the deployment of regional identities in the selling of these types of products, using the Dutch region of Waterland as a case study.

Research Context

The Case Study of Waterland

The example used here is part of a larger explorative study examining the production and reproduction of regional identities in the Netherlands in the period 1950 – 2000 (Simon, 2001; Simon et al., 2001). The region of Waterland is located just north of Amsterdam in the province of North-Holland (Figure 3.1), and is roughly enclosed by the *Markermeer* (eastern border), the city of Amsterdam (southern border), the *Zaanstreek* (western border) and *West-Friesland* (northern border). In the year 2000, approximately 164,000 people lived in the area with a population density of 600 inhabitants per square kilometre land, comparing with the average of 468 in the total of the Netherlands.[2]

As the name indicates, Waterland is a landscape full of water (according to the Dutch Central Bureau of Statistics 106 km² is water of a total area of 380 km²). Since the seventeenth century larger and smaller lakes were reclaimed, for example the *Beemster*, which was designated by UNESCO as a world heritage site in 1999. Often the region is described as an open peat-pasture landscape (Heidinga, 1977; De Pater et al., 1989) due to the occurrence of bog peat and the numerous drainage channels. These ditches cause a parcelling with a variable character. The region is dominated by agriculture, and specific characteristics, like type of soil, water balance and parcelling, make Waterland suitable for pastureland and (grazing) animal farming. In the year 2000, 632 farms had grazing animals, compared with 736 farms in total (Table 3.1).

Like other rural areas in Western Europe, the number of farmers is diminishing (from 942 in 1988 to 736 in 2000). The centres of the villages, in particular in the eastern part, are often demarcated as conservation areas. Furthermore, since the region is so closely situated near Amsterdam and many inhabitants work in the city, Waterland is also a residential area and a well-used recreation area for people from Amsterdam.

Table 3.1 Agricultural land use and livestock production in Waterland (2000)

Land use	km²	Livestock	n
Arable farming	19.94	Cattle	32,054
Pastureland	157.83	Pigs	3,447
Open horticulture	1.58	Poultry	14,762
Glass horticulture	0.15	Horses	920
Uncultivated land	0.45	Sheep and goats	53,756

Source: Dutch Central Bureau of Statistics (2002/2003)

Figure 3.1 Waterland

In 1994 the Ministry of Agriculture, Nature Management and Fisheries, assigned Waterland the status of a Valuable Cultural Landscape (*Waardevol Cultuurlandschap*). Accordingly, for five years more than one million euros a year were made available for projects mainly aiming at the renewal of agriculture. This caused an impulse of new projects, such as the development of regional products and agricultural nature conservation, by the inhabitants and political discourses.

Data Sources

The chapter is based on representations of Waterland in tourist folders and logos. Most 'regional actors' distribute information about their work and the area of work through folders (mainly directed at tourists). These play a dual role. First, they try to draw tourists to specific locations and attractions, by using images and words, which might idealize the site or attraction. Secondly, they reflect an image or representation of a place or region, providing visitors with narratives about the characteristics and meanings associated with the places concerned. Actors (here the developers of the material) bring forward those representations that they see themselves as symbols for a place or region. Through these representations, identities are constructed, which means that folders convey meaning to a region, and actors contribute to the production and reproduction of regional identities. Therefore, folders are a 'powerful narrative [...] of values and ideas' (Hopkins 1998, p. 65). As a result, folders are seen as commodities where regional identities are 'sold' with the purpose to attract more tourists and to inform potential visitors about attractions and characteristics of a region or place. The contents of ten folders that focus entirely on the region of Waterland are analysed in this chapter.

Like folders, logos also function as powerful symbols that construct identities, of the organization itself or the 'place' where the organization is located. Logos are graphical representations aiming to encourage 'consumption' and creating elements of recognition. As Hopkins (1998, p. 74) says:

> By creating a recognisable trademark, the commodity and/or its sponsors are given an identity or brand recognition which conveys positive, often multiple and abstract messages.

Based on their research on images of industrial cities, Barke and Harrop (1994) state that images and symbolism in logos influence the recognition of place identities. Hence, logos might be seen as products that use regional identities to sell perceptions. In this chapter focus is on representations of Waterland in the logos of regional organizations.

The research is also directed to the development of projects and activities. As Ilbery and Kneafsey (1998) state, such projects and activities may be signified through associations with particular places and regions. Consequently, they might be seen as 'regional products' or commodities, in this case of Waterland. Whereas there is growing interest in rural regional imagery and place promotion, little attention has been given to the promotion of quality products and services from specific regions (Ilbery and Kneafsey, 1998). This chapter tries to fill this gap, with special attention for the development of regional 'quality' products and educational material focusing on Waterland.

Furthermore, in January and February 2002, 13 in-depth interviews were conducted with representatives of regional organizations that are considered here as 'regional actors'. These organizations focus on tourists, the environment, farmers, rural development, water management, entrepreneurs and politics. The interviews aimed to collect data about 'who' ascribes 'which' identities to Waterland. Some questions are significant for this chapter, such as the importance and usage of identities in projects and promotional activities and the reasons why actors have chosen for this approach. Therefore the collected information is used as background material. The next section first describes Waterland as an imagined region in folders and Waterland as a trademark in logos. Then the analysis focuses on two activities developed in the region, Waterland as a quality product and Waterland as educational material.

Table 3.2   Representations of Waterland in folders

	Texts (%)	Photographs (%)
Environmental dimension	30	46
- Landscape	64	79
- Heritage	14	21
- Location	6	0
- Historical information	16	0
Economic-functional dimension	28	39
- Agriculture	45	38
- Recreation	32	31
- Nature conservation	20	13
- Living	0	13
- Remaining	2.5	5
Socio-cultural dimension	6	9
- Community	29	0
- Historical identities	35	0
- Traditions	36	100
Not used	36	7

Commodification of Waterland

Waterland as an Imagined Region

Three dimensions are used in categorizing representations of Waterland. The first concerns the question of the surroundings and environment of Waterland, or 'how Waterland looks'. The symbolic landscape of a region covers the natural and cultural landscape and heritage as defined by actors. It can be divided into four themes: landscape (wildlife, characteristics of the environment), heritage (for example, characteristics of historical buildings), location (references to the position, factors of location), and history. The second dimension is the economic-functional dimension, or 'how Waterland works'. This concerns activities related to the spatial use in Waterland, like agrarian, recreational and industrial activities, service industry, infrastructure, water management and nature conservation. The third dimension is the socio-cultural dimension, or 'what Waterland means as a way of living'. This includes the meanings, images, norms and values that are attached to Waterland, and tells something about the socio-cultural situation. The socio-cultural dimension is divided in three themes: the community (references to characteristics of the regional community); historical identities (myths, folklore and narratives); and traditions (regional languages, traditional customs, regional sports or festivities and activities aimed at this).

Through content analysis both sentences and photographs of ten tourist folders were classified according to these three dimensions to detect the dominant identities allocated to Waterland (Table 3.2). Each sentence or photograph may be classified in more than one dimension, and therefore, the score by dimension is higher than the total number of sentences and photographs. In total the folders contained 820 sentences and 119 photographs. As Table 3.2 shows, the environmental and economic-functional dimensions predominate the representations of Waterland. Identities constructed in tourist folders mainly focus on representations of characteristics of the landscape, and furthermore, Waterland is seen as an agricultural area with space for recreational activities and interest for nature conservation.

However, it was assumed that different actors would deploy different identities of Waterland in the selling of their products. Of the ten folders analysed, five were issued by actors in the recreational sector, one in the field of nature conservation and four in the agricultural field. Correspondingly, three different types of promotional folders were present. The first type of folders focuses on recreational activities in Waterland. The second type concentrates on characteristics of the nature and environment of Waterland, and these folders inform visitors about the flora and fauna in Waterland, the existence of different forms of landscapes and morphology, cultural-historical elements and the development of nature conservancy. As a third type, the folders selected involve a concentration on agriculture. Agriculture is seen as one of the main functions in Waterland, farmers being described both as the cultivators of agricultural products and also as conservators of the landscape. Figures 3.2 and 3.3 relate these types of folders with the representations of Waterland, classified in the selected dimensions.

The 'nature' folders confirm the assumption that the goals of organizations do correspond with the most used identities. It also fits for the 'agriculture' folders, although the photographs show a similar distribution in the functional and environmental characteristics that might be the result of interest in nature conservation by farmers in Waterland. The 'recreational' folders concentrate more on environmental characteristics and this has to do with the fact that the specific landscape of Waterland by itself is used as a pull-factor for tourists.

So far, representations of Waterland have been discussed from the perspective of regional identities in tourist folders. However, it is also interesting to look at representations of Waterland that refer to the past, the identities that are constructed as a sense of time. This can be done by counting sentences in the folders that focus on the history of Waterland or that emphasize the past in one way or another, for instance through the use of adverbs that relate to the past (Table 3.3). In the total number of 820 sentences, 112 refer to the past. All 112 sentences are classified in the environmental dimension with a division of 81 per cent referring to the history of Waterland and 19 per cent referring to the past through adverbs.

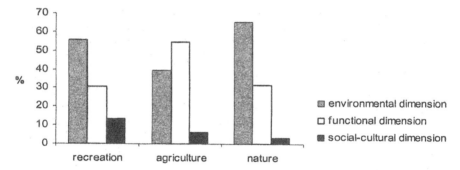

Figure 3.2 Relation between 'type' of folder and representations of Waterland in texts

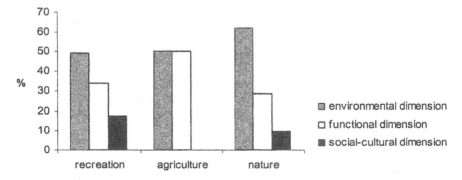

Figure 3.3 Relation between 'type' of folder and representations of Waterland in photographs

The high percentage (40 per cent) of sentences describing the influence of people in the forming of the landscape is striking. Focal points are the fight against water and the cultivation of the land. Representations of Waterland, seen as a construction of sense of time, mainly emphasize the fact that Waterland is a landscape made and constructed by people. So, the influence of water, directly via flooding and breaches in dykes and indirectly via the fight against water, is important in the history of Waterland, and this is clearly shown in the folders. In summary, representations of Waterland are often used in tourist folders and, therefore, they are elements in the process of commodification. With respect to the surroundings of Waterland, the representations of Waterland are dominated by the natural and cultural landscape. Apparently the landscape of Waterland is such a valued commodity that all 'regional actors' use it in the selling of Waterland as a tourist region.

Table 3.3 Selected quotations relating to Waterland

'In former ages, Waterland was even more watery than nowadays' (folder *Groot Waterland*)

'The tracks of the past are clearly visible in the landscape' (folder *Natuurlijk Waterland*)

'Sit down to the table of one of the restaurants and discover that the menu of Waterland is a pallet of tastes that reminds us of grandmother's kitchen' (folder *Waterland in Vogelvlucht*)

'This meat originates from the herbal grassland of Waterland, that causes the unique and old-fashioned nice taste' (folder *Waterlant's Weelde*)

'Through the age-long agrarian use, Waterland is a peat-pastureland of international meaning' (folder *Waterland Boerenland*)

'The landscape of Waterland is like a history book' (folder *Waterland Natuurlijk Boerenland*)

Waterland as a Trademark

Like folders, logos function as commodities because they can contain representations of regions. Logos of the 13 organizations interviewed were analysed, and next to the logos themselves, the underlying meanings are examined. Eight of the 13 logos include representations of Waterland, and comparable with the findings of Barke and Harrop (1994), the representations of Waterland mainly focus on characteristics of the landscape and architecture.

In three logos, the godwit is used as a marker. In Waterland, the number of meadow-nesting birds is so large that the godwit is almost seen as 'the' symbol of Waterland, and therefore, often used as an identity marker. Next to the symbol of the godwit, the Association of Agriculture and Nature Conservancy (*Vereniging Agrarisch Natuurbeheer Waterland*) tries to bring forward the farmer in its logo. 'Farmers and nature are bound together' is their motto, and through the use of this logo, attention is drawn to the importance of farmers in the conservation of the landscape. Two other organizations present Waterland more as images of nostalgia, familiarity and a rural idyll. The first organization, *Hotel de Boerenkamer*, offers overnight stays in 'authentic' farmhouses in Waterland. This image is clearly brought out in the logo by a drawing of a landscape with a typical farm of the area. In this way the organization tries to attract potential guests with the stimulus of an authentic feeling. *Waterlant's Weelde* is the second organization that uses a nostalgic logo. This organization sells regional quality products, and the representative claims that the logo stands for the bond with Waterland and that it creates an image which indicates that the product is tasteful and from perfect quality. The influence of this quality product is further explained in the next section. *Landschap Waterland*, the tourist organization, include in its logo, next to the shredding of parcels and reed, the numerousness of water. These symbols were chosen because, according to the interviewee, they are identifiable and expressive characteristics of the landscape of Waterland. The regional actors in Waterland use logos to promote their products and moreover, they use logos intensively as instruments to promote their region. In this way, they reproduce and reinforce the identities of Waterland.

Waterland as a 'Quality' Product

According to Bell and Valentine (1997), the production of food is important in indicating regional identities, or in other words, the development of regional quality products influences the commodification of regional identities in food products. A good illustration is given by Moran (1993) who showed, on the basis of the appellation of French wines, that the names of wines are often linked with region names. A strong link between product and place is seen because these 'wine producing regions' indicate a visible bond between the identity of the regions and the physical characteristics of regions. Bell and Valentine (1997, p. 155) state further that almost all products that correlate to a region can be sold through the embodiment of this region. In Waterland this is seen through the development of the regional 'quality' product sold under the name *Keurmerk Waterland* (or freely translated the 'control mark of Waterland').

The project started in 1996 to generate a surplus value for farmers. All the products relate to animal farming, like steak, lamb and cheese. The project was initiated with support from the Province of North-Holland thanks to the assignment of Waterland as a Valuable Cultural Landscape, and a stream of subsidies erected a whole new market line of regional products in Waterland. According to a key-informant of the *Stichting Keurmerk Waterland* four criteria are handled in the process of production. The first criterion is the gastronomic quality of the products, which involves the creation of an added value in quality and taste through the grazing on 'herbal' grassland. The key-informant stated it as follows:

> If you take care to conserve your grassland extensively, with the use of little artificial fertilizer, so that you have the guarantee that you have herbal grassland, than you make a contribution to an enormous effect in flavour in both dairy and meat products.

The second criterion is the quality of the production chain, with respect to the conservation of the grasslands of Waterland and animal rights. Farmers try not to spill fertilizer in surface water, the use of pesticides is minimized and the rules for animal treatment are similar to those applying to organic farming. Thirdly, the quality in the support of production is mentioned. All participating farmers are obliged with nature conservation on at least half of their land and here the *Keurmerk Waterland* goes even further than the 'eco-control mark'. The last criterion is the guarantee given to consumers that the products sold originated from Waterland. Farmers are residents from Waterland and cattle are raised in Waterland, and also the inputs in the production process originate from Waterland where possible. This criterion is linked with the wish of consumers to know the origin of products, which is the result of recent developments like mad cow disease, dioxin-chicken and genetically modified organisms. Quality becomes more important than quantity and the tracing of the place of origin completes this quality (Ilbery and Kneafsey, 2000).

Although the production of meat has no historical relationship with Waterland, it is commodified as typical of the region. The question is why these products are promoted as special for Waterland, or in other words, why is chosen for meat in a more or less 'dairy' farming region? Two reasons are conceivable. First, the production of meat was a practical choice because the production of milk as a regional quality product was not optional. Till recently, intervention in the dairy industry was hard because the three larger milk processing co-operatives did not believe in small-scale production and the above mentioned criteria. Second, meat is not directly bounded by quota, and for that reason, the production of meat was seen as a good alternative for dairy farms that generated insufficient income. The product is sold as it takes care for the preservation of grassland and open cultivated land with historical values. Next to the high quality standard, *Keurmerk Waterland* is thus commodified as a product that stands for the continuation of traditional farming in Waterland.

Educational Material: 'Looking for Waterland'

Another strategy for Waterland has been to enhance residents' knowledge of the characteristics of the region in the hope that this will increase support for preservation. In this section, special attention is given to the project, 'Looking for Waterland', the development of material for primary 'area' education because it aims to introduce pupils (4-12 years old) 'in a new, intense way with the landscape of Waterland, the land where they live, and therefore their responsibility with this landscape is growing' (Landschap Waterland, unknown). In 1995 the project started at the initiative of the regional tourist organization (*Landschap Waterland*), and in association with the schools and by support of subsidy due to the status of a Valuable Cultural Landscape. 'Area' education has been introduced into geography curriculum to increase the knowledge of the surrounding of a place or region, and the project 'Looking for Waterland' corresponds with this 'area' education. The project has three themes.

First, the pupils learn that meadow-nesting birds are important elements in the landscape of Waterland. This theme corresponds with the finding that birds are important identity markers in the commodification of Waterland in logos. The second theme explains the fight against water that is still continuing nowadays in the form of water management. Farmers want to have low water levels to improve their production and nature conservators want to keep the water level as high as possible to prevent the disappearance of peat and the corresponding landscape. The third theme covers stories of history and here the pupils also have to think about the future prospects of Waterland.

Through narratives, drawings, photographs, games, a video and a visit in the 'field', pupils learn to take responsibility for the conservation of the landscape of Waterland. In fact, the main product 'sold' in the project is, next to education, the landscape of Waterland. Pupils learn to care for their region and they learn how to preserve the landscape. Accordingly, the project aims to give attention to larger connections and relations in Waterland through the use of the natural and cultural landscape as commodities.

Conclusion

This chapter has described the use of regional identities by actors in the professional discourse. The case study of Waterland illustrates that regional identities are used in the process of commodification. Representations of Waterland are echoed in tourism folders, logos of regional actors and projects aimed at the preservation of identities. Apparently, regional identities are important elements in the process of commodification. However, the extent to which this take place varies between regions. Therefore, some assumptions are made why so many identities of Waterland are used in the commodification of products, and these assumptions may be applicable to other regions.

Recent projects and activities that implied the commodification of Waterland's identities were often related with the assignment of the status of Valuable Cultural Landscape by the national government. Because of this status subsidies were available for projects aiming to preserve the landscape in the present condition and to construct senses of place. Due to these projects regional identities of Waterland were often used in the 'selling' of products. Therefore, the first conclusion is that regional identities are more used in the process of commodification when there is enough money available.

However, money alone is not enough, since all projects and activities in Waterland would not exist without the effort from residents and other active actors. Because people are more aware of the uniqueness of their own surroundings, they want to care for 'their' region and therefore they disagree with the continuously increasing uniformity and large-scale projects. This is influenced by the increasing public interest in the quality of products due to recent problems in the agricultural sector and an overall increasing commitment with environmental issues. A second conclusion, therefore, is that when people develop a regional awareness because they are more interested in their 'place', identities of a region are more likely to be used as commodities.

Notes

[1] I would like to thank Paulus Huigen and Peter Groote for their useful comments, and Tamara Kaspers-Westra for her cartographic work.

[2] All numbers are based on data of eight municipalities (Beemster, Edam-Volendam, Landsmeer, Oostzaan, Purmerend, Waterland, Wormerland, Zeevang) given by the Dutch Central Bureau of Statistics 2002/2003.

References

Allen, J., Massey, D. and Cochrane, A. (1998), *Rethinking the Region*, Routledge, London.

Barke, M. and Harrop, K. (1994), 'Selling the Industrial Town: Identity, Image and Illusion', in J.R. Gold, and S.V. Ward (eds), *Place Promotion: the Use of Publicity and Marketing to Sell Towns and Regions*, John Wiley, Chichester, pp. 93-114.

Bell, D. and Valentine, G. (1997), *Consuming Geographies*, Routledge, London.

Bessière, J. (1998), 'Local Development and Heritage: Traditional Food and Cuisine as Tourist Attractions in Rural Areas', *Sociologia Ruralis*, **38**, pp. 21-34.

Brouwer, R. (1999), *Toerisme in de Arena*. PhD thesis, Wageningen.

Cloke, P., Goodwin, M. and Milbourne, P. (1998), 'Cultural Change and Conflict in Rural Wales: Competing Constructs of Identity', *Environment and Planning A*, **30**, pp. 463-80.

Graham, B. (ed.) (1998), *Modern Europe: Place, Culture and Identity*, Arnold, London.

Groote, P., Huigen, P.P.P. and Haartsen, T. (2000), 'Claiming Rural Identities', in Haartsen, T., Groote, P. and Huigen, P.P.P. (2000), *Claiming Rural Identities*, Van Gorcum, Assen, pp. 1-7.

Halewood, C. and Hannam, K. (2001), 'Viking Heritage Tourism: Authenticity and Commodification', *Annals of Tourism Research*, **28**, pp. 565-80.

Hall, S. (1997), *Representation: Cultural Representations and Signifying Practices*, Sage, London.

Heidinga, H.A. (1977), *Historie en Archeologie van Waterland*, I.P.P. Publicatie, 215.
Hoekveld, G. (1993), 'Regional Identity as the Product of Regional Integration', in E. Dirven, J. Groenewegen and S. van Hoof (eds), *Stuck in the Region? Changing Scales for Regional Identity*, VUGS, Utrecht, pp. 15-38.
Holloway, L. and P. Hubbard (2001), *People and Place: the Extraordinary Geographies of Everyday Life*, Pearson Education, Harlow.
Hopkins, J. (1998), 'Signs of the Post-rural: Marketing Myths of a Symbolic Countryside', *Geografiska Annaler*, **80B**, pp. 65-81.
Ilbery, B. and Kneafsey, M. (1998), 'Product and Place: Promoting Quality Products and Services in the Lagging Rural Regions of the European Union', *European Urban and Regional Studies*, **5**, pp. 329-41.
Ilbery, B. and Kneafsey, M. (2000), 'Producer Constructions of Quality in Regional Specialty Food Production: A Case Study from South-west England', *Journal of Rural Studies*, **16**, pp. 217-30.
Jackson, J.B. (1995), 'A Sense of Place, a Sense of Time', *Design Quarterly*, **164**, pp. 24-6.
Johnston, R.J., Gregory, D., Pratt, G. and Watts, M. (eds) (2000), *The Dictionary of Human Geography, Fourth Edition*, Blackwell, Oxford.
Jones, O. (1995), 'Lay Discourses of the Rural: Developments and Implications for Rural Studies', *Journal of Rural Studies*, **11**, pp. 35-49.
Kneafsey, M. (2000), 'Tourism, Place Identities and Social Relations in the European Rural Periphery', *European Urban and Regional Studies*, **7**, pp. 35-50.
Kneafsey, M. (2001), 'Rural Cultural Economy: Tourism and Social Relations', *Annals of Tourism Research*, **28**, pp. 762-83.
Knox, P. L. and Marston, S.A. (2001), *Places and Regions in Global Context*, Prentice Hall, Upper Saddle River, New Jersey.
Landschap Waterland (unknown), *Op Zoek in Waterland*, Landschap Waterland, Monnickendam.
Markwick, M. (2001), 'Marketing Myths and the Cultural Commodification of Ireland: Where the Grass is Always Greener', *Geography*, **86**, pp. 37-49.
Mitchell, C.J.A. (1998), 'Entrepreneurialism, Commodification and Creative Destruction: a Model of Post-modern Community Development', *Journal of Rural Studies*, **14**, pp. 273-86.
Moran, W. (1993), 'The Wine Appellation as Territory in France and California', *Annals of the Association of American Geographers*, **83**, pp. 694-717.
Murphy, A.B. (1991), 'Regions as Social Constructs: the Gap between Theory and Practice', *Progress in Human Geography*, **15**, pp. 22-35.
Pater, B.C. de, Hoekveld, G.A. and van Ginkel, J.A. (1989), *Nederland in Delen, een Regionale Geografie. Deel I: Nederland als Geheel, West- en Zuidwest Nederland*, De Haan, Houten.
Pellenbarg, P. H. (1991), *Identiteit, Imago en Economische Ontwikkeling van Regios*, Geo Pers, Groningen.
Ray, C. (1998), 'Culture, Intellectual Property and Territorial Rural Development', *Sociologia Ruralis*, **38**, pp. 3-20.
Simon, C. (2001), 'Streekidentiteit: Modegril of Blijvertje?', in: F. Daalhuizen and S. Heins (eds), *Rurale Diversiteit en Dynamiek. Een Wetenschappelijke Visie op het Nederlandse Platteland in de 21e Eeuw*, Bergdrukkerij, Amersfoort, pp. 79-90.

Simon, C., P. Groote and P. Huigen (2001), 'Verstreking of Ontstreking?', *Rooilijn*, **1**, pp. 16-22.

Vlieger, J.J., van Ittersum, K. and van der Meulen, H.S. (1999), *Streekproducten: van Consument tot Producent*, Lei, Den Haag. Rapport 3.99.13.

Chapter 4

'That Quintessential Repository of Collective Memory': Identity, Locality and the Townland in Northern Ireland

Bryonie Reid

Introduction

It is clear that remembered and current geographical insecurities play a significant part in the Northern Irish conflict. There are severe problems in imagining the six counties as a coherent place; it fits neatly neither into the dominant visions of Ireland as a whole, nor those of Britain, and each of these larger entities is itself increasingly fragmented. The ways in which both unionism and nationalism traditionally construct their claims to the territory of Northern Ireland are centred on absolute ownership and mutual exclusion. Unionist and Protestant narratives of belonging on the land have been couched in terms of suffering and hard work, while nationalist and Catholic narratives have focused on prior settlement and use. This chapter aims to uncover alternative means of establishing belonging in place in Northern Ireland. Nuances exist, suspended between and around the simplified narratives; taking into account the ongoing failure of the concept of 'nation' to provide an inclusive or complex ground for belonging in Northern Ireland, the chapter focuses on the potential of the locality to do so. Rather than arguing for localities as inherently inclusive of cultural, political and religious diversity in themselves, the concern is with inclusive *uses* and *understandings* of localized space. Unfortunately, many people in Northern Ireland do not regard inclusion as being a positive; although these uses of locality could be defined as progressive, that view is not universally shared among unionists and nationalists in the North. It seems obvious, however, that some kind of shared belonging needs to be negotiated between the two communities laying claim to the six counties, despite their evident reluctance to do so.

The chapter begins by indicating schemes for a Northern Ireland which would be able to transcend national struggles, by being imagined as less than, and more than, a nation. This includes thinking on Northern Ireland as one region of Europe, detached from its national loyalties; as common ground for both nationalist and unionist, different from Britain and Ireland and thus providing shared as well as contested space; and as a 'cultural corridor' between Britain and Ireland (Longley, 1991, p. 144). A review of the intensely localized nature of the Troubles in Northern Ireland follows, as a challenge to any easy theorizing of the local as a solution to the province's opposing belongings.

The argument develops with reference to the case study, the more positive example of the recent use of townlands. Historically, the townland has been a fundamental element of understanding rural and urban space in Ireland, although its place in the address system remained essential only in rural areas. In 1972 the Post Office decided that the townland element of the address was obsolete in Northern Ireland, and would be replaced by house numbers, road names and postcodes. In response, the Townlands Campaign evolve to resist this disruption of centuries of identity and belonging in place, and the concomitant assault on local repositories of memory. The campaign, a ground-level community effort, apparently bypassed the usual frictions and clashes of such issues in Northern Ireland. This is a rare and encouraging example of unity, and of a focus on place and belonging that moves beyond traditional nationalist or unionist concerns. It leads to the contention that uses of Northern Irish localities need not necessarily demonstrate narrow and exclusive definitions of place, belonging and identity, and conflicting remembering, but, conversely, may provide a significant counterpoint to the rigid and contradictory manifestations of those concepts at the national level. The issue of the eradication of townlands, and the resulting widespread protest, exists in a detailed and ecumenical pattern of belonging woven from diverse localized threads.

The Parish and the Cosmos: Finding Local Belonging in Northern Ireland

A common analysis of the current global situation, if any analysis pretends to such a broad remit, is that while on the one hand, the world is becoming homogenized and ever more accessible to certain groups, on the other, there is a growing concern with the local, the familiar and the distinctive. It is possible to read this latter phenomenon as a response to the former. Increased mobility (forced or voluntary), competition between various cultures, and the standardizing impulse of global capitalism, may all combine to create a sense of dislocation, loss and anxiety, rather than of freedom, progress and choice. An at times nostalgic focus on place is closely related to the relatively recent vogue for memory work, another corollary of these changes. Pierre Nora identifies the phenomenon as the loss of something sacred, where memory has become isolated, archival, and discontinuous, rather than being as it once was, 'integrated . . . all-powerful, sweeping, unselfconscious and inherently present-minded' (Nora, 1996, p. 2). He also imagines memory as 'something rooted in the soil' (Nora, 1996, p. 11), making explicit the perceived interdependence of remembering and belonging in place. The growing concern with memory work, manifested in archives, museums, history theme parks, heritage tourism and television, is considered by some commentators to be a symptom of anxiety over: identity loss and disorientation in place; the much-vaunted 'end of history'; and the 'postmodern condition' (Evans and Lunn, 1997, p. vii). Arguably, therefore, a need exists to preserve difference and to establish belonging in community and place at a local level.

Addressing the dichotomy and necessary dialogue between the global and the local in Ireland, Kearney muses on the possibility of a 'fifth province', envisaged as supplement and other to Ireland's existing four, and located at 'the swinging door which connects the "parish" . . . with the "cosmos"' (Kearney, 1997, p. 100). This concern with opening out attachment to place takes into account the apparent need to be rooted somewhere, but warns against what Ascherson terms a 'black, locked up, excluding sort of provincialism' (Ascherson, 1990, p. 16). Kearney's notion of the fifth province evolves specifically within the framework of a partitioned Ireland, torn by battles to control and define nationhood. Europe, with its long history of racial violence, and its more recent attempts at integration, has been spoken of in terms of regions rather than of nations, a conceptualization that serves to soften the rigidity of borders, and the associated conflicts across the continent. A regionalized Europe has been welcomed by Kearney as an opportunity to dissipate the Northern Irish conflict by decentralizing Britain and Ireland, Northern Ireland then becoming one of a 'quasi-autonomous, albeit interconnected' group of regions in the Western European archipelago (Kearney, 1997, p. 93).

Northern Irish nationalist politician, John Hume, interprets the drive for European unity as a force compatible with both nationalist and unionist aspirations for the province; nationalists may 'take comfort from the fact that an ever closer union applies to both parts of Ireland within Europe', while unionists may focus on 'an ever closer union . . . between Britain and Ireland' in the same framework (Hume, 1996, p. 126). Pointing out that, in this new Europe of 'multiple' peripheries, and 'diverse' centres (p. 131), nationalism and unionism are somewhat anachronistic concepts, Hume contends that an emphasis on Northern Ireland's wider connections could help dissipate its parochial conflict. These theories found an earlier advocate in Ulster poet John Hewitt, the descendant of English Protestant planters. He attempted to respond to the evident need for discourses of belonging in Northern Ireland that would avoid 'a political mystique of Irishness' (Longley, 1992, p. 19), and provide more cultural depth and authenticity than the traditional British connection.

Proposing that Catholic and Protestant could at least consider Northern Ireland shared ground, and base a sense of belonging on their common geographical heritage, Hewitt contextualized that attachment within a wider network of belonging on the island of Ireland, in Britain and in Europe. His democratic vision of belonging was founded on time spent in place and the accumulation of place-based memory, rather than race or religion:

> for we have rights drawn from the soil and sky;
> the use, the pace, the patient years of labour,
> the rain against the lips, the changing light,
> the heavy clay-sucked stride, have altered us;
> we would be strangers in the Capitol;
> this is our country also, nowhere else;
> and we shall not be outcast upon the world
> (Hewitt, 1991, p. 79).

A further attempt to envisage Northern Ireland as a shared and distinctive space occurs in Edna Longley's well-known evocation of the province as a 'cultural corridor' (Longley, 1991, p. 144), in which 'Irishness' and 'Britishness' permeate and alter one another. She pins hope on a culturally hybrid regionalism in literature and other cultural expression, situated within at least two broader contexts (Britain and Ireland); there are echoes here of Bhabha's (1994) work on the subversive potential of the hybrid, in terms of both space and person. He considers that borderline situations and permeable, mongrel identifications critique implicitly the purifying drive of nationalisms and the homogenizing force of global capitalism. As Longley (1991, p. 144) acknowledges, however, unionism and nationalism tend to 'block the corridor at one end . . . [or] the other'. Similarly, Hewitt later declared his notion of the distinct and shared space of Northern Ireland a failure, conceding: 'Ulster is not one region, it's several regions . . . My concept of regionalism was trying to bring together incompatible pieces' (cited in McDonald, 1995, p. 45). Hewitt's theories stumbled over Northern Ireland's real and minutely scaled geographical differentiation, and likewise, Bhabha's theories present certain problems when contextualized in actuality. Mitchell (1997) believes that Bhabha's premise almost re-essentializes liminal spaces and characters, in that he supposes them *inevitably* to enshrine resistance. Those criticisms may be applied to any suggestion that localities are subversive in themselves rather than in their uses. Mitchell argues that borderline spaces, given their existence in 'real' geography and history, could as easily become spaces of 'closure and cultural homogenization' as has clearly occurred in Northern Ireland (Mitchell, 1997, p. 537).

'That Determined Place': The Challenge of Sectarian Geography (McDonald, 1995, p. 58)

The presentation of the local as an alternative to national allegiances (and hence conflicts) in Northern Ireland is seductive. However, the elevation of the local as a space that transcends constitutional questions and essentialisms faces a glaring difficulty, which is that all too often, it does nothing of the kind. Northern Ireland may be described as an intermingled, borderland area, but this context has resulted more often in fossilization of identity and difference than any kind of acceptance, let alone celebration of, ambiguity. This is evident not least in the realm of remembering. Smith (1999) identifies the memorial underpinnings of Irish nationalism as memories of pagan and Christian golden ages, the claim to the ancestral lands, and histories of heroic self-sacrifice and suffering. By contrast, unionism in Ulster enshrines:

> vivid shared memories of the vicissitudes of the first settlements, and memories of earlier golden ages after the Battle of the Boyne and the siege of Londonderry . . . coupled with a deep-rooted belief in ethnic election as God's covenanted people . . . [producing] a powerful attachment to Ulster as the ancestral and promised homeland of the Protestant settlers (Smith, 1999, p. 273).

The sectarian exclusions of memory historically extend to geography, and the last thirty years in particular have been characterized by further polarization of settlement. Areas that may at first glance seem to be mixed often simply break down into more and more intricate patterns of segregation, which are reinforced time and again in response to violence. A. T. Q. Stewart sums up the problem as follows:

> the two communities are not intermingled . . . but they are interlocked . . . This gives rise to a situation in which the "territorial imperative" is extremely insistent . . . the war in Ulster is being fought out on a narrower ground than even the most impatient observer might imagine, a ground every inch of which has its own associations and special meaning . . . locality and history are welded together (Stewart, 1977, pp. 181-2).

Communities have painfully learned the risks of living alongside the 'other'; McKay (2000) cites several cases of sectarian murder involving Catholics and Protestants from the same locality, using this situation to question unreflective tendencies to value localized integration. Territorial localization in Belfast is now such that distrust between neighbouring Protestant communities matches that between adjoining Protestant and Catholic areas. David Holloway indicates in an essay on the Protestant Donegall Pass in Belfast that the boundaries are now being drawn at a smaller scale than ever:

> I was confronted with a community that had a clear sense of geographical separation and a strong identity [*sic*] from all the surrounding communities, in terms of rivalry, antagonism and mistrust, regardless of religion. For the people of the Donegall Pass, there was no blurring of the boundaries (Holloway, 1994, p. 9).

Belonging within national discourses is fraught with contradiction and characterized by simplification and selective amnesia. Belonging within local discourses can be equally oppositional, fragmented and traumatic, yet contains the potential to complicate sectarian models. An unassuming example of this potential occurs in the Townlands Campaign, providing a surprising counterpoint to the usual political acrimony. Faced with the threat of townland names and boundaries disappearing from use and memory in Northern Ireland, Protestants and Catholics have marshalled consistent and voluble protest, both separately and together. The spaces and their names, abundant in their associations with local identity and belonging, seem to be considered largely as a shared resource and heritage. In its working from the ground up, so to speak, the campaign provides a litmus test for theories of local inclusion, a prism which acts to reflect and deconstruct abstract conjecture.

The Townlands Campaign: 'The Child's Open-eyed Attention to the Small and the Familiar' (Heaney, 1980, p. 142)

The History of the Townlands Campaign

The townland is one of the most basic units of land division in Ireland, 'the most intimate and enduring' (Ó Dalaigh et al., 1998, p. 9). The first official evidence of their existence occurs in church records from before the twelfth century (Muhr, 1999, p. 5) and, thereafter, they appear in all legal documents pertaining to land usage and ownership. As a sanctioned form of understanding Irish land division and naming, most of these spaces and names were preserved practically untouched throughout the vagaries of the plantation years. Across Ireland the townland retained its historical function in administrative and legal circles, its position confirmed as the principal element of rural addresses, only for this to be terminated abruptly in 1972 by the Post Office's now current system of using road names, numbers, and postcodes. None the less, many communities deemed road names and numbers and postcodes unacceptable. One road with a single name may cut through several townlands, and road names could be assigned with no reference to local names and in ignorance of local opinions and knowledge. Townland names were not banned, but were considered 'superfluous information' (Kirk-Smith, 1993, p. 46), and people were requested not to include them on addresses. It may be argued that significant layers of localized memory were being erased in this initiative, and objections were voiced almost unanimously across the province. The names, resonant in their associations with local identity, memory and belonging, seem to have been considered as a shared resource and heritage. However, county councils were the government bodies responsible for validating such a change, and as the structures of local government were undergoing alterations at this time, the Post Office's directive was 'allowed . . . to become law, almost by default' (McCool, 1993). Fermanagh is the only county which managed to resist the scheme completely.

In the mid-1970s, a campaign to rescind the Post Office's guidelines was launched through the Federation for Ulster Local Studies, an umbrella organization for various community groups concerned with local studies. The Federation perceived a need to coordinate a response and reassure people of their townland's validity; if the names fell into disuse, they and the exact spaces to which they referred would certainly be forgotten. The initial reaction was one of outrage that a structure of names and spaces that had such a lengthy history in the community was to be discarded for the Post Office's convenience. As Kay Muhr points out, 'the townland was for centuries the building block of local society' (Muhr, 1999, p. 3), and it came as a shock to have those building blocks of identity and belonging summarily abandoned. Although the Post Office lacked the power to make townland names illegal, its discouragement of their use gave some rural dwellers that impression; anxious that cheques and other official post should be able to reach them, the Post Office's directive was often followed despite protest. The Federation's first statement on the matter, published in its journal in 1976, warned that:

> the Post Office system, where implemented, is leading to the disuse of the townland name, firstly by official bodies, and then by members of the public who are made to feel that there is something improper or unreliable about its use (Federation for Ulster Local Studies, 1976, p. 21).

The Meaning of the Townlands Campaign

The evidently strong desire among communities in Northern Ireland to protect local distinctiveness from the globalizing imperative is important. Townlands, while not exceptional in the sense that similar small land divisions would have been known elsewhere in Europe, are uniquely long-lived. This framework, an integral part of spatialized identity in Ireland for almost a millennium, is a prime example of the geographical scale in which E. Estyn Evans considers 'the most genuine bonds' to form between person and place (Evans, 1967, p. 8). These small, 'pragmatic' units (Ó Dalaigh et al., 1998, p. 10), are redolent of an attachment to place which is both crumbling and petrifying in the wake of increasing homogenization. None the less, the understanding of identity and belonging that emanates from townland-measured orientation in place can be both modest and far-reaching. The attack on this form of local grounding has elicited numerous protests: to cite but one:

> the townland is at the heart of rural thought . . . the loss will be not only of a sonorous beauty and historical sense but also a corruption of the intimate relationship between rural people and the land that surrounds them (McCool, 1993, p. 8).

Kirk-Smith (1993, p. 46) ascribes 'psychological and moral' meanings to the townland in this context. Rather than providing the impulse for a merely narrow and nostalgic celebration of the local, 'while a tidal wave of international culture sweeps inexorably across the land, conquering and absorbing the local, the townland has been at the centre of a campaign of resistance'. The simple act of paying attention to 'the eclectic and the ordinary, . . . [to] local particularity in a context of universal process', a quality of Evans' work in which Graham (1994, p. 198) finds contemporary resonance, can perhaps act in itself as an antidote both to the destruction and the exploitation of a sense of belonging in place. Seamus Heaney too has lauded the power of the local:

> empowered within its own horizon, it looks out but does not necessarily look up to the metropolitan centres . . . it is self-sufficient but not self-absorbed, capable of thought: undaunted, pristine, spontaneous: a corrective to the inflations of nationalism and the cringe of provincialism (in Canavan, 1991, p. xi).

Indeed, it may form the foundation of a more gentle, careful, and free-ranging thinking than does a rootless universal perspective. Nash (1999, p. 471), for example, recognizes and questions the 'tendency to celebrate hybridity over authenticity, dislocation over location, mobility over rootedness'.

The potential of townlands to resist homogenizing bureaucracy extends further. An early objection to road names and numbers that involved the disruption of an ancient understanding of place in Ireland is put succinctly by one writer; 'a townland is an area, it is not a long thin thing called a road' (in McGurk, 1993, p. 8). In 1977 the Federation for Ulster Local Studies stated: 'we feel [the Post Office scheme] is destroying the pattern of knowledge associated with the townland system' (Federation for Ulster Local Studies, 1977, p. 24). This pattern of knowledge and memory is spatially oriented, as distinct to the linearity of postal addresses. In reference to Northern Irish space, the replacement of the townland system with the road system echoes early Elizabethan administrative strategies in Ireland. Hadfield and Maley describe early modern Ireland as 'both a mirror and hammer' in relation to English identity; while its colonizers and governors attempted to define themselves by moulding Ireland into their Other, its 'complex, differentiated, heterogeneous and variegated' nature repeatedly shattered that image (Bradshaw et al., 1993, p. 15 and p. 3). The response focused on persistent attempts at control through mapping, where land and inhabitants could be textually ordered and categorized. It has been observed, however, that as the Irish often only recognized English borders in order to weight their incursion with meaning, 'colonial 'plotting' embodies just as much the disturbance of its own categories as their establishment . . . it declares itself a ceaselessly violated invention' (Baker, 1993, p. 88).

It could be argued that a similar process is occurring in the imposition of linear geography over an already existing spatial mode of orientation; this 'thick' space (see Herr, 1996) entails the possibility of resistance to reorganization from above, as thoroughly enmeshed as it has been with human life and death. In the context of the limited and inflexible definitions of belonging in place and community in Northern Ireland, townlands and their names represent the possibility that the locality involves multifaceted and dishevelled stories of memory, identity and belonging, although these may not necessarily be recovered; they also bear the distinction of lengthy use and association. Ultimately, road names have proved unsatisfactory. They have not 'evolve[d] naturally from within the community' (Turner, 1977, p. 26), and inadequately represent local memories. The townland names were held to symbolize generations of memories rooted in places, a narrative of the intimate and mingled nature of that remembering.

Central to the Townlands Campaign's importance in this context is its apparent ability to garner support across communities. The literature makes frequent reference to this characteristic, often mentioned almost as an aside rather than as the Campaign's main ethos. Hence there are comments such as: 'support for our position has been virtually unanimous and has come from many different kinds of people' (Turner, 1977, p. 26), and 'townlands belong to us all . . . they are part of our shared heritage' (Federation for Ulster Local Studies, undated). In several places it is mentioned that the issue of preserving the townland system proved to be a uniting force for Fermanagh District Council, unusually enough in its history, and elsewhere it is noted that the idea to include townland names on road signs apparently originated with an Orangeman (see Carroll, 1996; Kirk-Smith, 1993; Muhr, 2002). It seems that the origins of townlands and their names may not be a matter of contention or even importance to many of the campaigners; for them the value of the spaces lies in their protracted use and human associations, in their links with the past, and in their stories about landscape and settlers. Of course, townlands are as vulnerable to assimilation into conventional religio-political geographies as any other space. An example of this occurs in South Armagh, in the parish of Forkhill. There, the Mullaghbane Community Association has erected granite markers on townland boundaries, signposting the names in Anglicized form and Roman letters, and below in Gaelic form and script. These efforts to preserve local land divisions simultaneously provide a layer of visual differentiation as 'Irish' for the area, and the area's positioning on the border between Northern Ireland and the Republic cannot help but intensify the political nuances of the markers.

Nash writes of the Townlands Campaign as a refreshing rethinking of attachment to place in the Northern Irish context, as people unite against the 'common threat' (Nash, 1999, p. 469) of bureaucratized nomenclature and construction of place. If local knowledges can supplant the discourses of ethnicity and religion when speaking of belonging in place, the notion could be opened out, a step which the Townlands Campaign seems to be making at times, whether consciously or not. This process is resonant of Hewitt's understanding of belonging, in which time creates attachment to place and allows roots to be put down, eventually making no distinction between planter and native. As Seamus Heaney writes:

> the associations of the word [townland] are rural, of course, but I suspect that its talismanic power is felt by city people also . . . it connotes a totally uninsistent sense of difference, a freely espoused relation to an idiom and an identity that are regional, authentic, uncoerced and acknowledged. It is a minimal but reliable shared possession, the kind of word that could provide the right verbal foundation for talks about talks (in Canavan, 1991, p. xi).

Despite the largely shared interest in townlands in rural areas, and although Heaney reminds us that townlands as spatial divisions exist in cities also, the localized territorial patterns of urban conflict warn against placing too much hope in the reconciliatory potential of local attachment. Belfast's streets and townlands continue to be violently and rigorously segregated, clearly evident in the fierce conflict in 2001 between the neighbouring areas of the Short Strand and the Lower Newtownards Road. Perhaps the most to be said for the use of townlands in this context is that they are at least a common way of understanding Belfast's divided geography.

However, the overwhelming tenor of the interest in townlands has been inclusive and preservative in nature, the emphasis being on an existing mongrel heritage rather than the creation or recovery of a pure one. There is also no constitutional question in the balance, whatever the outcome of the Campaign. It seems to have survived largely unco-opted by unionism or nationalism, working beneath and beyond their more elevated and more limited concerns. The failure of the local to open out in urban areas is a serious drawback to generalized theory on its progressive possibilities. The power of the Townlands Campaign, where it exists in positive form, is something indefinable, ordinary, unassuming and easily missed, something sufficiently fragile that, if made to bear the burdens of 'Peace' or 'Identity', it might crumble. A grand solution to the geographical attritions of Northern Ireland seems unlikely, but a beginning can be made through the use of townlands in observing, drawing attention to, and participating in the rich web of names and spaces we inherit from the various past.

Conclusion: 'The Anthology of Memories of the Other is a Book I Hadn't Reckoned on . . .' (Ian Crichton Smith, in Falconer, 1998, p. 13)

This web of spaces spatially and imaginatively divides Northern Ireland into individual localities at various scales. At some level, many have a stated political, cultural or religious identity, no matter their actual or historical diversity. This situation emanates from the lack of an overarching, embracing narrative of place to which all inhabitants of Northern Ireland may subscribe, and in part explains how it has failed conceptually as a geographical entity in its own right. Seamus Heaney highlights the difficulties of such competing constructions of place-identity as do exist:

> the fountainhead of the Unionist's myth springs in the Crown of England but he has to hold his own on the island of Ireland. The fountainhead of the Nationalist's myth lies in the idea of an integral Ireland, but he too lives in exile from his ideal place (Heaney, 1984, p. 5).

This dual geographic understanding of Northern Ireland has been an integral part of its conflict, creating 'the strain of being in two places at once' (Heaney, 1984, p. 5). There have been various attempts to resolve or to dissolve the problem. Resolutions have often been couched in terms of being able to establish prior possession and hence authenticity; the nationalist and republican movements have until recently gone unchallenged in their use of the Irish-Ireland myths. Now some unionists and loyalists are beginning to reciprocate, creating Ulster-Scots myths to vindicate their presence in Northern Ireland through prehistory, language and the naming of place; this constitutes an attempt to fix Protestant belonging in the north of Ireland as 'immemorial and uncontrollable' (Foster, 1988, p. 78).

That there are deep problems of belonging in place and remembering in Northern Ireland cannot and should not be denied. There are, however, different stories that emerge quietly beneath the close look, which enable a kind of remembering that questions established boundaries and oppositions. Localities investigated on an intimate scale will reveal selective and sectarian modes of remembering, belonging, and identifying, but there will also exist counter-memories, belongings, and identities. Beneath the coarse political certainties lies a finer, more variegated pattern, which encompasses the larger structure, but is not defined by it. Rather than the 'abstractly conceived whole, the cosmos . . . the rationalistic preoccupation with the universal' (Plumwood, cited in Davion, 1994, p. 15), the particular should be attended to, provided its scrutiny is open to the possibility of disharmony at that scale as at all others. A focus on particularity or specificity of place allows the recording of more detailed and more complex experiences. Thus Loughrey (1986, p. 211) imagines townland names as 'the index-cards upon which memories were stored', possibly a more hopeful scale than a Northern Ireland of 'cynical, selective forgetting' or 'responsible, alarming memory' (Longley, 2001, p. 253).

Belonging to place is an integral part of Northern Ireland's flimsily constructed polarizations. The townland in itself has no inherently inclusive qualities; indeed, it could be argued that localities are inherently *ex*clusive through the very act of defining them as one place and not another. However, the interpretations to which townlands have been subjected through the Townlands Campaign demonstrate an awareness of the complex texture of local spaces, a texture that can make partisan simplification difficult. This is not to claim that the Townlands Campaign is the key to achieving peace; its importance here lies in the sidestepping of such universals.

Hewitt's work on Ulster's forgotten histories can be compared to that of a farmer on an overgrown field, 'clearing away the undergrowth of lies and amnesia . . . as groundwork for the future, a way of clearing weeds and encouraging healthy growth' (Clyde, 1987, p. viii). This analogy recognizes the need for patient and persistent exposure of Northern Ireland's real historical multiplicity, which might be distinguished more effectively at the level of the local, although local studies do not inevitably deconstruct the dominant religious, ethnic, and political narratives. For this reason the local should not be theorized as a solution to Northern Ireland's sectarian conflict. Its uses, as evidenced by the Townlands Campaign, may possibly form, rather, a clearing, a reversed palimpsest, a 'thick' space in which certainties are complicated, and things forgotten aired. Thus it can be argued that to root historical variety in geographical locality is to provide the common environment that permits dialogue between forms of belonging. Rooted and tangled memories linger in local places, shaping identity and belonging; perhaps obscured, tarnished through disuse or grief, but awaiting careful excavation, examination and reinstatement in the record of what has gone before.

References

Ascherson, N. (June 1990), 'The Four Motors are Driving Off', *Fortnight*, **296**, pp. 16-19.
Baker, D. (1993), 'Off the Map: Charting Uncertainty in Renaissance Ireland', in B. Bradshaw, A. Hadfield, and W. Maley (eds), *Representing Ireland: Literature and the Origins of Conflict, 1534-1660*, Cambridge University Press, Cambridge, pp. 76-92.
Bhabha, H. (1994), *The Location of Culture*, Routledge, London.
Bradshaw, B., Hadfield, A. and Maley, W. (eds) (1993), *Representing Ireland: Literature and the Origins of Conflict, 1534-1660*, Cambridge University Press, Cambridge.
Canavan, T. (ed.) (1991), *Every Stoney Acre has a Name: A Celebration of the Townland in Ulster*, Federation for Ulster Local Studies, Belfast.
Carroll, R. (1996), 'Townlands they Love so Well', *Irish News*, 8th April.
Clyde, T. (ed.) (1987), *Ancestral Voices: The Selected Prose of John Hewitt*, Blackstaff Press, Belfast.
Davion, V. (1994), 'Is Ecofeminism Feminist?', in K. Warren (ed.), *Ecological Feminism*, Routledge, London, pp. 8-28.
Evans, E. E. (1967), 'The Irishness of the Irish', paper given at the Annual Gathering of the Irish Association, Armagh, 22nd September.
Evans, M. and Lunn, K. (eds) (1997), *War and Memory in the Twentieth Century*, Berg, Oxford.
Falconer, A. (1998), 'Remembering', in A. Falconer and J. Liechty (eds), *Reconciling Memories*, Columba Press, Dublin, pp. 11-19.

Federation for Ulster Local Studies (1976), 'The Post Office and Rural Addresses in Northern Ireland – a Federation Statement', *Ulster Local Studies*, **2(1)**, pp. 21-23.

Federation for Ulster Local Studies (1977), 'Federation News: Secretary's Report, 1976-1977', *Ulster Local Studies*, **3(1)**, pp. 23-26.

Federation for Ulster Local Studies (undated), 'Townlands: a User's Guide', informal publicity material.

Foster, R. (1988), *Modern Ireland: 1600-1972*, Penguin, London.

Graham, B. (1994), 'No Place of the Mind: Contested Protestant Representations of Ulster', *Ecumene*, **1**, pp. 257-81.

Heaney, S. (1980), *Preoccupations: Selected Prose 1968-1978*, Faber and Faber, London.

Heaney, S. (1984), *Place and Displacement: Recent Poetry of Northern Ireland*, Trustees of Dove Cottage, Grasmere.

Herr, C. (1996), *Critical Regionalism and Cultural Studies: From Ireland to the American Mid-West*, University Press of Florida, Gaineville.

Hewitt, J. (1991), *The Collected Poems of John Hewitt*, Blackstaff Press, Belfast.

Holloway, D. (1994), 'Territorial Aspects, Cultural Identity: the Protestant Community of Donegall Pass, Belfast', *Causeway*, **Winter**, pp. 9-12.

Hume, J. (1996), *Personal Views: Politics, Peace and Reconciliation in Ireland*, Town House, Dublin.

Kearney, R. (1997), *Postnationalist Ireland: Politics, Culture, Philosophy*, Routledge, London.

Kirk-Smith, I. (March 1993), 'Going Local', *Fortnight*, **315**, p. 46.

Longley, E. (1991), 'Opening Up: a New Pluralism', in R. Johnstone and R. Wilson (eds), *Troubled Times: Fortnight Magazine and the Troubles in Northern Ireland, 1970-1991*, Blackstaff Press, Belfast, pp. 141-44.

Longley, E. (1992), 'Writing, Revisionism and Grass-Seed: Literary Mythologies in Ireland', in J. Lundy and A. Mac Póilin (eds), *Styles of Belonging: The Cultural Identities of Ulster*, Lagan Press, Belfast, pp. 11-21.

Longley, E. (2001), 'Northern Ireland: Commemoration, Elegy, Forgetting', in I. McBride, (ed.), *History and Memory in Modern Ireland*, Cambridge University Press, Cambridge, pp. 223-53.

Loughrey, P. (1986), 'Communal Identity in Rural Northern Ireland', *Ulster Local Studies*, **Autumn**, pp. 205-11.

McCool, J. (1993), 'Last Chance to Save Townland', *Belfast Newsletter*, 9th February, p. 8.

McDonald, P. (1995), 'The Fate of "Identity": John Hewitt, W. R. Rodgers, and Louis MacNeice', in E. Patten (ed.), *Returning to Ourselves: the Second Volume of Papers from the John Hewitt International Summer School*, Lagan Press, Belfast, pp. 41-60.

McGurk, J. (1993), 'Crusade to Save Ancient Names of Townlands Gathers Strength', *Irish News*, 4th March.

McKay, S. (2000) *Northern Protestants: An Unsettled People*, Blackstaff Press, Belfast.

Mitchell, K. (1997), 'Different Diasporas and the Hype of Hybridity', *Environment and Planning D: Society and Space*, **15**, pp. 533-53.

Muhr, K. (1999), 'Celebrating Ulster's Townlands: A Place-Name Exhibition for the Millennium', Ulster Place-Name Society, Belfast.

Muhr, K. (2002), 'Townland News: A Report on the Work of the Ulster Place Name Society', *Due North*, **1(5)**, pp. 16-18.

Nash, C. (1999), 'Irish Place Names: Postcolonial Locations', *Transactions Institute of*

British Geographers, **24**, pp. 457-81.

Nora, P. (1996), *Realms of Memory: The Construction of the French Past*, Columbia University Press, New York.

Ó Dalaigh, B., Connell, P. and Cronin, D. (eds) (1998), *Irish Townlands: Essays in Local History*, Four Courts Press, Dublin.

Sharkey, S. (undated), 'Of Salt, Song, Stone and Marrow-bone: The Work of Anne Tallentire', in V. Connor (ed.), *Anne Tallentire* (exhibition catalogue), Project Press, Dublin, pp. 25-40.

Smith, A. D. (1999), *Myths and Memories of the Nation*, Oxford University Press, Oxford.

Stewart, A.T.Q. (1977), *The Narrow Ground: Aspects of Ulster 1609-1969*, Faber and Faber, London.

Turner, B. (1977), 'The Post Office and Rural Addresses', *Ulster Local Studies*, **2(2)**, pp. 26-7.

Chapter 5

Mapping Meanings in the Cultural Landscape

Yvonne Whelan

Introduction

> The city ... does not tell its past, but contains it like the lines of a hand,
> written in the corners of the streets, the gratings of the windows, the
> banisters of the steps, the antennae of the lightning rods, the poles of the
> flags, every segment marked in turn with scratches, indentations, scrolls
> (Calvina, 1974, p. 41).

The re-emergence of cultural geography in the 1980s paved the way for a renewed appreciation of the symbolic spaces that prevail in the built environment. Cognisant of the highly textured and multi-layered nature of landscape, geographers began to conceive of it as a site of emblematic representation, a constructed space, comprising elements that play a pivotal role in the construction, mobilization, and representation of identity. From the names attached to city streets, to the national parliament building of a capital city, the symbolic dimensions of the cultural landscape are everywhere apparent, signifiers of identity and memory, and tangible sites of cultural heritage. Perhaps the iconography of landscape is most especially obvious in the case of grand, empire building projects and cities designed under the dominance of dictatorial control. Think, for example, of the ways in which cities of Eastern Europe were moulded in the image of socialist leaders to express power and control and also to cultivate a sense of national identity (Åman, 1992). Symbolic spaces, however, are not just confined to such audacious empire-building projects. On the contrary, as Denis Cosgrove points out, 'All landscapes are symbolic ... reproducing cultural norms and establishing the values of dominant groups across all of a society' (Cosgrove, 1989, p. 125). In Northern Ireland, for example, annual parades harness the symbolic power of space through the careful routing of marches. First and Second World War memorials in local towns and rural villages are invariably topped by a lone male figure, armed with a rifle.

Unofficial monuments, murals and memorials commemorate the victims of the Troubles, while kerbstones are variously painted red, white and blue or green, white and gold. The appropriation of public spaces in this fashion transforms otherwise innocuous features of the cultural landscape into highly charged symbolic sites.

This chapter explores the ways in which particular aspects of the cultural landscape, chief among them public statues, street names, architecture and urban design initiatives, serve as significant sources for unravelling the geographies of political and cultural identity. It is argued that the dynamic relationship between history and geography is demonstrated when, for example, national monuments, public buildings and streets celebrating national heritage are inserted into the landscape in a manner that maps history onto territory. These landscape elements act as spatializations of power and memory, making tangible specific narratives of nationhood and reducing otherwise fluid histories into sanitized, concretized myths that anchor the projection of national identity onto physical territory (Bell, 1999). As Kong argues:

> landscapes are ideological in that they can be used to legitimize and/or challenge social and political control (Kong, 1993, p. 24).

Recent research in the field draws upon an analytical framework borrowed from art history, the iconographic method, to reveal and interrogate the values and meanings embedded in the visual imagery under consideration (Rose, 2001). This approach also draws upon notions of landscape as text, that is, a social and cultural document which can be read in order to reveal the many layers of meaning and processes written into it and the ways in which these change through time (Duncan and Ley, 1993). Such approaches ultimately reflect the polyvocality of present-day societies and acknowledge implicitly that a single landscape may be subject to a range of interpretations depending on the cultural and political position of the interpreter. They also point to the fact that cultural landscapes are complex sites of meaning, formed out of the set of social relations and practices which prevail and interact at any one particular location and time. This point is explored towards the close of this chapter through a case study analysis of the changing nature of Dublin City's iconography as it moved from being a city of empire to an independent capital.

Interrogating Icons of Identity in the Cultural Landscape

> Symbols are what unite and divide people. Symbols give us our identity, our self-image, our way of explaining ourselves to others. Symbols in turn determine the kinds of stories we tell; and the stories we tell determine the kind of history we make and remake (Mary Robinson, Inauguration

speech as President of Ireland, December 3, 1990).

The cultural landscape, together with the signs and symbols that comprise it, plays a crucial role in legitimating particular political and social orders and in contributing to narratives of group identity. Where places evolve in contentious political circumstances and make turbulent transitions, for example, from the colonial to the post-colonial, particular aspects of landscape take on special significance. In different ways and to varying extents they play a crucial role in creating the iconography of landscape. These 'icons of identity' also draw upon the cultural capital of the past in order to reinforce the dominance of particular ruling authorities, while at the same time they can acts as focal points around which resistance and opposition can be channelled, especially in the context of post-independent cities when they become implicated in strategies of nation-building.

Carving Memory in the Built Environment

Among the most strikingly symbolic features of any town, city or urban landscape are the public monuments which line its streets and dot its squares. Commemorating an individual or an historic event, public monuments are not merely ornamental features of the urban landscape but rather highly symbolic signifiers that confer meaning on the city and transform otherwise neutral places into ideologically charged spaces. While the ancient Greeks used them as a means of conferring honour on esteemed members of society, it was not until the mid-nineteenth century that public statues took on particular significance as a means of celebrating a nation's past. Up to the outbreak of World War One, the statue served as a symbolic device of enormous popularity, a means of imposing the ideals and aspirations onto the public consciousness in a way that other cultural signifiers could not (see Johnson, 1995). The intense nationalism of these years gave rise to widespread and sustained attempts across Europe and North America to commemorate national histories in monumental form as:

> statues sprouted up on the public thoroughfares of London at a rate of one every four months during Victoria's reign. The streets and public squares of even modest-sized German towns bristled with patriotic sculpture: in a single decade some five hundred memorial towers were raised to Bismarck alone (Owens, 1994, p. 103).

The frenzy of monument building that occurred in Europe stemmed largely from the fact that national governments recognized their key role as foci for collective participation in the politics and public life of villages, towns and cities. Statues served to strengthen support for established regimes, instilled a sense of political unity and cultivated national identity. As Lerner observes:

embedded within the monument is a particular way of staging politics that is centred on the spectacle or visual display. With its emphasis on representing human forms, the monument reveals two important terrains upon which political power and the form of the nation rest: the spectacle of politics and the public display of the body. Like the bronze and stone figures themselves, these terrains have tremendous resilience over time (Lerner, 1993, p. 178).

Thus monuments that previously been confined to the private domain spread into the arena of public, secular space and increasingly came to dot the squares, principal thoroughfares and public buildings of cities, town and villages.

For those concerned with understanding the dynamics at work in shaping the historical and the contemporary urban landscape, acts of memorialization are of much significance. The objects of a people's national pilgrimage, monuments are signifiers of memory which commemorate events or individuals but also:

> mark deeper, more enduring claims upon a national past as part of the present... monuments may become both historical symbols of nationhood and fixed points in our contemporary landscapes (Withers, 1996, p. 327).

As Sandercock suggests, we erect sculptures to dead leaders, war heroes and revolutionaries because:

> memory, both individual and collective, is deeply important to us. It locates us as part of something bigger than our individual existences, perhaps makes us seem less insignificant... Memory locates us, as part of a family history, as part of a tribe or community, as a part of city-building and nation-making (Sandercock, 1988, p. 207).

If the city is a repository of collective memory, then public statues make an important contribution to its memory bank while focusing attention on specific places and events in highly condensed, fixed and tangible sites.

As dynamic sites of meaning and memory that transform neutral spaces into sites of ideology, public statues not only help to legitimate structures of authority and dominance but are also used to challenge and resist such structures and to cultivate alternative narratives of identity. Just as public statues served throughout the late nineteenth and early twentieth centuries as a means of cultivating popular support, making power concrete in the landscape, the medium was also employed, paradoxically, by groups at odds with established regimes as a means of challenging the legitimacy of governments and objectifying the ideals of revolutionary movements. For the very qualities that make public statues so valuable in building popular support for established regimes also make them a useful target for those who wish to overthrow such regimes. This is particularly the case in post-colonial countries emerging from beneath the shadow of the colonial enterprise or ideological domination, the urban landscapes of cities in Eastern Europe being a case in point. In the wake of the fall of communism the dissolution of the 'Iron Curtain' precipitated the mass removal of a vast number of public statues dedicated to the monolithic figures of communist rule such as Marx, Stalin and Lenin. More recently, the end of Saddam Hussein's rule in Iraq in 2003 was signalled when monuments of the dictator were triumphally toppled throughout the land.

Naming Places, Claiming Spaces

Another important means by which societies, both today and in the past have utilized the past as a cultural, political and social resource is through the naming of towns, cities and, in particular, streets. Affixing names to places is inextricably linked with nation-building and state formation and consequently sweeping changes in the naming process reflect ideological upheavals. For example, the city of St. Petersburg, so named in the era of Tsarist Russia, was renamed Leningrad under communism and in a gesture of much symbolism, reverted back to St. Petersburg in a referendum in June 1991. In both colonial and post-colonial contexts, the names given to streets take on great symbolic significance in tandem with the obvious functional significance of a street name. Just as public statues are central to the creation of national mythologies and the invention of traditions, so too street names, albeit in a different, perhaps less audacious way, manage to commemorate past events and figures and are used by political regimes in order to legitimate and consolidate their dominance and reinforce their authority. Traditionally, street names tended to be vernacular in origin, designating geographical orientation and urban function and often drawing upon elements of local topography and history. This began to change in the eighteenth century, however, when, with the introduction of postal services, the responsibility for naming fell into the hands of officialdom. Gradually street names began to take on greater political significance, a trend which gathered pace in the latter half of the nineteenth century, when naming became a prominent feature of the age of modern nationalism, colonialism and empire building.

As dominant powers set about cultivating a sense of collective identity and creating a shared past, official and authorized versions of history were often made concrete in the urban cultural landscape though the use of street names. Naming streets after famous persons or events perpetuated in the streetscape and in the minds of inhabitants the memory of historical figures or events deemed worthy of remembering by those in charge. In effect:

> commemorative street names, like their alphanumerical counterparts, provide locational information, but they also have the function of perpetuating, reifying and constructing a view of the past (Azaryahu, 1996, p. 311).

The selection of street names, therefore, is an inherently political procedure determined by ideological needs and political power relations and always implemented by agents of ruling power. Street names conflate history and geography and bring the past that they commemorate into the everyday language and ordinary settings of human life, thereby transforming history into a feature of the natural order of things (Azaryahu, 1997). Naming plays a crucial role in processes of empire building, as well as in the rituals of revolution and rebellion that mark the emergence of postcolonial nations when divesting the namescape of past historical associations becomes important. Any discussion of naming, therefore, must also consider the phenomenon of renaming and the role of names in challenging the legitimacy of historical traditions and in building national identity. Street names can serve as focal points around which expressions of dissent and opposition can be anchored. Citizens often demonstrate their opposition to those in power by resisting the names imposed by a higher authority, while re-naming streets, just like destroying monuments, has an immediate effect on the politics of landscape and serves as an act of political propaganda of enormous symbolic value. Thus, in both everyday conversation and in literary narratives, place-names have a semantic depth that extends beyond the concern with simple reference to location or to a single image (Entrikin, 1991).

Building and Designing the Urban Landscape

Just as street naming and public statuary are crucial in the construction of identity narratives, so too are architectural and urban design initiatives. The politically motivated construction and reconstruction of urban spaces has long fascinated geographers interested in exploring the interrelations between ideology and landscape. Many urban landscapes have been laid out, if not at their very beginning, then at some point in their existence, according to a plan. This morphological framework, often accompanied by a written text, is also bound up with the politics of power and identity. As Kenny argues:

> a planning document, possibly more than any other written text, articulates the ideology of dominant groups in the production of the built environment (Kenny, 1992, p. 176).

For centuries the city form has been planned and manipulated in order to represent power and ideology. Features such as squares and road patterns, whether axial, orthogonal or gridiron, contribute to a geometry that is radically different from the curves and undulations of the natural landscape but which represent human reason and the *power* of intellect (Cosgrove, 1989). Countless examples of civic design initiatives the world over demonstrate the fact that the city plan, the skeleton around which an urban landscape develops, is an orchestrated piece of work with much symbolic underpinning, and which plays a particularly important role in representing discourses of power. This includes both the plans that made the transition from the drawing board to reality, and also those that remained ideas. For example, cities such as Washington D.C., New Delhi or Canberra, the European cities like Berlin, Moscow and Rome that were re-shaped during the totalitarian regimes of the 1930s or cities of North Africa shaped during colonial rule each embody in their planned layout a considerable measure of symbolic potency.

Within the structures of the city plan, the specific significance of architecture demands attention. More than merely assemblages of bricks and mortar, buildings are invested with meaning and possess a particular representational and aesthetic value alongside their inherently practical function. As Jencks suggests of architecture:

> not only does it express the values (and land values) of a society, but also its ideologies, hopes, fears, religion, social structure, metaphysics (Jencks, 1982, p. 178).

Buildings effectively communicate by symbol the culture of which they are a part, representing that culture as well as housing it in a manner that draws upon the symbolic capital of the built form. It is, therefore, one of the geographer's tasks in reading cities, to uncover the meaning and politics invested in the buildings which prevail in the cultural landscape. In interpreting such buildings it is important to situate them amid the cultural and political context in which they were built. Issues of location, architectural style, function, design, ornamentation and iconography, the type of building material employed and the ceremony and ritual associated with their construction and opening, along with how their function changes over the course of time, are all aspects which become important when interrogating the iconography of the built environment. Public buildings provide an important means of gaining access to the meaning embedded in the urban landscape.

Reading Dublin: Iconography and the Politics of Identity

In concluding this discussion of the symbolic nature of the cultural landscape, it is worth exploring in more detail one case study example which underscores the significance of landscape as a site of emblematic representation. Over the course of centuries, the fortunes of Dublin City have waxed and waned in tandem with broader political developments. The trajectories of Irish politics, the complicated relationship with London, and Dublin's rather ambiguous status as a city of empire, were to place significant demands on its symbolic fabric, and in particular, the monumental landscape. Before 1922, the island of Ireland was firmly incorporated into the United Kingdom of Britain and Ireland. It became the canvas upon which the British administration, and agents loyal to it, set out to paint a picture of union and loyalty to empire. Indeed, the five visits of three different royal monarchs to the island in the first 11 years of the twentieth century clearly demonstrated this. Towns and villages throughout the land were lavishly decorated and parade routes were carefully plotted. Space became politicized in a manner that underscored the island's status as a constituent component of empire. Ireland was also, however, as the historian Stephen Howe points out, 'a sphere of ambiguity, tension, transition, hybridity, between national and imperial spheres' (Howe, 2000, p. 68). The implications of this contested political context for the monumental spaces of the island were many, and can be explored through the lens of the iconography of Dublin's cultural landscape at the turn of the twentieth century.

Before the achievement of political independence the urban landscape became highly politicized and contested. While in some respects landscape became a means of cultivating a sense of imperial identity and fostering a feeling of belonging to empire among citizens, it also served as a site upon which to represent resistance (Whelan, 2002; 2003). Then a deposed capital, the city stood at an important geographical interface or contact zone between the imperial metropole and the colonized territory. In 1900, for example, visitors to the city would have found themselves in a landscape peopled with a variety of figures cast in stone. Statues that had been erected during the eighteenth and nineteenth centuries commemorated a myriad of figures closely connected to the British administration, from kings and queens, to members of the military establishment. In the heart of the city, Kings William III, George I and George II, Lord Nelson and the Duke of Wellington each occupied dominant positions. These monuments were representative of a broader trend that prevailed throughout the country to embody in stone the link with empire. Points of physical and ideological orientation, they comprised one dimension of a monumental landscape that was consistently augmented over the course of three centuries. Towards the end of the nineteenth century, however, concerted attempts were made to subvert the inscription of imperial power and to challenge the legitimacy of British rule. Monuments were erected to heroes of the nationalist movement, while attempts were made to rename streets and new building projects were planned which would project in stone the desire for political independence.

Political developments in the early years of the twentieth century paved the way for the emergence of the Irish Free State, a process that was punctuated by the Easter Rising of 1916, the War of Independence, and the Civil War. As Dublin became capital of the Free State, it played a significant role in the nation-building process. Although the leaders of the first generation after Independence evoked an image of Irish society that was almost exclusively rural, various aspects of the urban landscape did play an important role in marking the transition from the colonial to the postcolonial. What were once the linchpins in both the visual expression of imperial rule and part of a strategy of resistance to the colonial power became instead essential tools in supporting the ideology of the new regime. The powerful symbolism of architecture and urban planning initiatives was recognized in the early decades of the fledgling state when particular attention was directed towards the proposed building of a national parliament complex while, at the same time, the rebuilding of vast swathes of destroyed urban landscape provided further opportunities to harness the symbolic power of public space. Equally significant was the unveiling of additional monuments in the capital after 1922. The new administration, just like the previous regime, recognized the powerful role of the past in legitimating authority and forging identity. It was, however, a very different past that was celebrated and commemorated in Dublin after 1922. While old relics of empire were summarily removed, new heroes of the independence struggle were to be erected on plinths throughout the capital and country. Both commemoration and de-commemoration, therefore, proved central to the new discourses of power and identity that were mapped out in the post-Independence period.

Thus in Dublin before and after the achievement of political independence, symbolic spaces proved to be especially important in underpinning narratives of national identity. With increasing distance from the independence struggle, however, and the inevitable cultural maturing of the State, it would seem that such spaces no longer retain the powerful significance they once possessed. Monuments that had been erected with such choreographed ritual as symbols of a nationalist ideology would seem to have lost much of their symbolic potency in the contemporary context. This raises some interesting issues regarding the contemporary iconography of the island's space. In many ways the new 'Spire of Dublin' speaks volumes of the cultural and political climate that now prevails in Ireland. As the centrepiece of O'Connell Street's integrated action plan, it is intended to become the signature of the city in the 21st century, a Dublin version of Paris's Eiffel Tower or Sydney's Opera House. This monument is striking for a number of reasons, not least its utter simplicity, its ahistorical nature and complete lack of any political association. These characteristics combined to make it, in the eyes of the judging panel, 'an ideal emblem for the current time.' In these respects the 'Monument of Light' stands in marked contrast to earlier emblems of nationhood that marked the Free State's symbolic geography after 1922. It captures in microcosm a much broader, paradigmatic shift that has taken place in Irish political and cultural life during the modern period. It is a shift from dependence to independence, from the modern to the postmodern and from being an inward looking island on the periphery of Europe, to an outward looking contributor to the European super-state in an era of globalization. It symbolizes the changing conception of culture that now prevails and which transcends narrow ethnic boundaries. This conception recognizes that Irish identity crosses over the simple binary oppositions of Catholic and Protestant, nationalist and unionist, republican and loyalist and is contested along new axes of differentiation.

Conclusion

The symbolic approach to understanding cultural landscapes underscores the fact that they are highly complex discourse in which a whole range of economic, political, social and cultural issues are encoded and negotiated (Daniels, 1993). As Seymour argues:

> the view comes from somewhere, and both the organization of landscapes on the ground, and in their representations, are and have been often tied to particular relationships of power between people (Seymour, 2000, p. 194).

The powerful role of public monuments, street naming and architectural and urban design initiatives in shaping particular narratives of identity across a variety of different spatial scales and in an equally broad range of political, social and cultural contexts has been the subject of a burgeoning body of research in historical and cultural geography. This work demonstrates that signs and symbols are integral in giving concrete representation to forces of dominance as well as of resistance and opposition. Although central to processes of empire formation and nation-building, landscape represents more than the impress of state power or elite ideologies. Rather the cultural landscape embodies *many* interwoven layers of power and overlaps with issues of race, gender, class and local identity politics. Cultural and historical geographical research on the symbolic meanings of landscapes has therefore attempted to disclose the dynamic nature of the relationship between landscape, memory and power in the construction of nation identity. It has been shown that landscapes, both representational and material act as a means of articulating discourses and identities and as a medium for the representation of competing discourses among different social and political groups. Moreover, landscape iconographies can be read and interpreted as symbolic representations of larger power struggles between competing interests and identities. They underscore the fact that every landscape is:

> a synthesis of charisma and context, a text which may be read to reveal the force of dominant ideas and prevailing practices, as well as the idiosyncrasies of a particular author (Duncan and Ley, 1993, p. 329).

References

Åman, A (1992), *Architecture and Ideology in Eastern Europe during the Stalin Era*, MIT Press, London.

Azaryahu, M. (1996), 'The Power of Commemorative Street Names', *Environment and Planning D: Society and Space*, **14**, pp. 311-30.

Azaryahu, M. (1997), 'German Reunification and the Politics of Street Names: The Case of Berlin', *Political Geography*, **16**, pp. 479-93.

Bell, J. (1999), 'Redefining National Identity in Uzbekistan', *Ecumene*, **6**, pp. 183-207.

Berg, L. D. and Kearns, R.A. (1996), 'Naming as Norming: "Race", Gender, and the Identity Politics of Naming Places in Aoteraoa/New Zealand', *Environment and Planning D: Society and Space*, **14**, pp. 99-122.

Calvino, I. (1979), *Invisible Cities*, Picador, London.

Cosgrove, D. (1989), 'Geography is Everywhere: Culture and Symbolism in Human Landscapes', in D. Gregory and R. Walford (eds), *Horizons in Human Geography*, Palgrave, Basingstoke, pp. 118-35.

Daniels, S. (1993), *Fields of Vision: Landscape Imagery and National Identity in England and the United States*, Polity Press, Cambridge.

Duncan, J. and Ley, D. (eds) (1993), *Place/Culture/Representation*, Routledge, London.

Entrikin, J.N. (1991), *The Betweeness of Place*, John Hopkins University Press, Baltimore.

Howe, S. (2000), *Ireland and Empire*, Oxford University Press, Oxford.

Johnson, N. (1995), 'Cast in Stone: Monuments, Geography and Nationalism', *Environment and Planning D: Society and Space*, **13**, pp. 51-65.

Kenny, J. (1992), 'Portland's Comprehensive Plan as Text: The Fred Meyer Case and the Politics of Reading', in T.J. Barnes and J.S. Duncan (eds) (1992), *Writing Worlds: Discourse, Text and Metaphor in the Representation of Landscape*, Routledge, London, pp. 176-92.

Jencks, C. (1982), *Architecture Today*, Doubleday, New York.

Kong, L. (1993), 'Political Symbolism of Religious Building in Singapore', *Environment and Planning D: Society and Space*, **11**, pp. 23-45.

Lerner, A.J. (1993), 'The Nineteenth-Century Monument and the Embodiment of National Time', in M. Ringrose and A.J. Lerner (eds), *Reimagining the Nation*, Open University Press, Buckingham, pp. 176-96.

Owens, G. (1994), 'Nationalist Monuments in Ireland, 1870-1914: Symbolism and Ritual', in R. Gillespie and B.P. Kennedy (eds), *Ireland: Art into History*, Townhouse, Dublin, pp. 103-17.

Rose, G. (2001), *Visual Methodologies*, Sage, London.

Sandercock, L. (1988), *Towards Cosmopolis*, John Wiley, New York.

Seymour, S. (2000), 'Historical Geographies of Landscape', in B. Graham and C. Nash (eds), *Modern Historical Geographies*, Pearson, Harlow, pp. 193-217.

Whelan, Y. (2002), 'The Construction and Destruction of a Colonial Landscape: Monuments to British Monarchs in Dublin Before and After Independence', *Journal of Historical Geography*, **28**, pp. 508-33.

Whelan, Y. (2003), *Reinventing Modern Dublin. Streetscape, Iconography and the Politics of Identity*, University College Dublin Press, Dublin.

Withers, C.W.J. (1996), 'Place, Memory, Monument: Memorialising the Past in Contemporary Highland Scotland', *Ecumene*, **3**, pp. 325-44.

Chapter 6

Exploring the Irish Mumming Tradition with GIS

Amanda McMullan

Introduction

This chapter is concerned with methodological developments for a particular expression of intangible cultural heritage, namely folklore or ethnology. Ethnological elements are discovered at a certain time, amongst a certain group of people and in specific places. Therefore, ethnological research methodologies, which ignore the spatial dimension of phenomena, may be excluding valuable new insights, interpretations and understandings. As a computer-based tool that emphasizes the spatial dimension, GIS offers many functions for turning data into information to assist in our understanding of the phenomena being studied. Although developed originally for the study of environmental issues, the number of GIS applications has been expanding into new and diverse areas. However, 'cultural, historical, and other such geographers, who have not taken quantitative approaches in the past, are now beginning to use GIS' (Murayama, 2001, p. 165). Nevertheless, the cultural domain has generally been resistant to the methodology and, despite evidence that it might hold considerable promise, it has not been widely used as a tool for research into cultural heritage. There is, however, increasing interest in the application of GIS to the interrogation of historical data, particularly in establishing trends and changes over time and across space. Such developments can be attributed to the realization that, in addition to its mapping functions, GIS can be a flexible and valuable tool which can assist with many aspects of research including fieldwork, data management, and data processing. But the acceptance or otherwise of the application of a GIS methodology to a research project also involves a range of social factors that are unrelated to the technical limitations of the technology (Campbell and Masser, 1995). Thus the attitudes and interpersonal skills of the researchers involved in a particular GIS application can play an important role in determining its acceptance and overall success (Reeve and Petch, 1999). This is especially relevant, given that researchers working in the cultural domain typically come from disciplines hitherto resistant to GIS technology. This chapter examines the application of a GIS to an Irish ethnological case study entitled 'Room to Rhyme', which is a collaborative research project investigating the historical, sociological and geographical aspects of the mumming tradition in Ireland. (The partners include: Department of Irish Folklore, University College Dublin; Academy for Irish Cultural Heritages, University of Ulster; and the Ulster Folk and Transport Museum.) Mumming is a form of folk drama in which, traditionally, a dozen or more performers visited neighbouring homes in the weeks before Christmas and performed a short play. The play's core is a mock hero-combat scene, around which the relatively short dialogue revolves. The heroes in opposition are, generally, Prince George and St. Patrick (or the 'Turkish Champion') one of which is slain and then revived by a 'doctor'. A succession of performers then enters onto the stage to recite a series of rhymes. These include characters with names such as 'Father Christmas', 'Beelzebub', 'Johnny Funny', or historical figures like 'Oliver Cromwell', or more exotic figures such as 'Jack Straw' and the 'Green Knight'. More recently, the traditional house visit has been replaced by performances in public houses and community halls, and the raising of funds for charitable projects has superseded the object of collecting money for the purposes of 'festive cheer'.

'Room to Rhyme' involves disparate forms of qualitative data, which are often considered unsuitable for a GIS and largely involves researchers who have had no previous experience in GIS methodology. The case study, however, requires detailed spatial analysis which makes it a useful example through which to explore the potential for using GIS in ethnological research. Following a contextual discussion that locates the project within the overall context of the validity of GIS methodology to research in cultural heritage, the chapter examines its particular application to the 'Room to Rhyme' project. Finally, this application is evaluated within the context of the wider problems attending the integration of GIS and ethnological research.

Context

GIS and Cultural Research

In general terms, current research practices increasingly encourage the use of multiple methods in a single study, as the strengths of one method may compensate for the limitations of another (Philip, 1998). Normally, GIS is considered as part of an array of quantitative/spatial analytical methods and, consequently, is usually distinguished from the qualitative methods more widely employed in cultural research including folklore and ethnology. Nevertheless, there are exceptions. For example, feminist researchers have taken a lead in exploring the potential for GIS in cultural research (see for instance: Kwan, 2002; McLafferty, 2002; Pavlovskaya, 2002; Schurmann and Pratt, 2002). In so doing, they emphasize that such integration of research methods can enrich feminist research while also extending the possibilities of the GIS methodology. Further, while relatively few in number, the conclusions of case studies applying GIS to other fields of cultural research suggest that the technology can enhance and complement existing research methodologies in cultural research. Two examples serve to illustrate this point.

First, Dow (1994) examines the potential for GIS as a research tool in cultural anthropology. He observes that an important problem for this discipline lies in the difficulty of locating cultural groups and examining their relationship to natural and human environments. Dow explored the ability of GIS to resolve this problem by employing the technology in a study that sought to map language in a rural area of Mexico. He argues that the inclusion of GIS caused the entire project to progress more quickly than previous studies. In particular, he notes that the GIS could process data and produce effective maps more quickly and inexpensively in the field compared to traditional methods. More importantly, the GIS could merge together diverse sources of information to produce high-resolution maps. These small-scale cultural maps made it possible to 'look at where people live on a scale that is small enough to reveal the features of their environment to which their cultures respond' (Dow, 1994, p. 479). In addition, the high-resolution GIS maps revealed new geographical distributions, which differed greatly from those generated in previous studies. For example, earlier studies produced only rough paper maps of the locations of the different linguistic groups. These indicated that an ethnic group, the Tepehua, was geographically dispersed. The high-resolution GIS cultural maps revealed, however, that the ethnic group had two distinct areas of concentration. Dow concludes that the strength of employing GIS in cultural anthropology is that it can solve old problems, while allowing the field to retain the strength of its fieldwork principles by maintaining close contact with its informants and the culture that it is examining. What is more, Dow claims, the mapping procedures used in the study are applicable to many types of field research in the social sciences.

Secondly, in a study of the social ecology of the residential patterns of members of immigrant churches in Houston, Texas, Ebaugh et al. (2000) used GIS to map 11 immigrant congregations and the residential addresses of their members. This data was combined with the results of ethnographic fieldwork in order to understand how social ecological variables influenced the organizational characteristics of religious institutions. Ebaugh et al. conclude that the use of GIS technology enriched the study of immigrant congregations in various ways. Primarily, it provided both statistical and visual data, which enabled them to determine accurately the geographical dispersion of congregation members, and the distances that they lived from their place of worship. This information alone, however, was insufficient to explain the structures of congregations. Consequently, Ebaugh et al. argue that although GIS mapping technology can provide valuable data, these results are only suggestive of the motivations, desires, interests and investments which members have in their particular congregation. Only observational and interview data can probe the nuances of meaning which members ascribe to their membership and involvement in their congregations (Ebaugh et al., 2000, p. 115).

Thus they conclude that the *combination* of GIS techniques and traditional field methods of social research has the potential to maximize understanding of research problems, particularly those with spatial ramifications.

GIS and Mumming

The mumming tradition in Ireland is precisely one such issue. Mumming is associated essentially with areas of predominantly English settlement during the seventeenth-century plantations of Ireland. These include certain districts on the east coast as well as large swathes of Ulster where colonization was most intense. While the mummers' play was once the preserve of English settlers in Ireland, it was quickly embraced by Gaelic Irish as well as by Scottish planters. 'Room to Rhyme' was concerned with documenting and recording the current survival of mumming in the north of Ireland, where the tradition is evidently strongest and where local variation in text, character composition and performance is most marked. In addition, 'Room to Rhyme' seeks to explore variations in the tradition through the geographical analysis of different aspects of the plays, which may illuminate the factors that have shaped the nature of the mumming tradition at different locations in the North of Ireland. The integration of a GIS into this project also provides the potential to produce a more nuanced spatial understanding of mumming than could be obtained from conventional ethnological mapping.

According to Gailey (1972) ethnological research has three elements; historical; sociological; and geographical. He argues that the latter is fundamentally important to analysis since 'cultural elements are discovered not just at a certain time and amongst a certain group of people, but also at a specific place' (Gailey, 1972, p. 121). Indeed, it can be argued strongly that ethnological research is incomplete if it disregards the geographical dimension to its subject matter. It is this cardinal importance of the spatial, and the ways in which spatially distributed phenomena change through time, which makes ethnological research such a suitable arena for the application of GIS methodology. Moreover, GIS offers significant advantages over conventional mapping, which has always been an important analytical tool for ethnological research. The creation of a visually effective map, which is also an important component of interpretation, is impeded by the static nature of the paper map, which makes it difficult, expensive and time-consuming to make changes to the finished product. As a result, 'paper map' cartographers have very little scope for trial and error in the creation of their maps and one of the original motivations for the development of GIS was actually the elimination of such problems. Nevertheless, paper mapping continues to be the conventional cartographic methodology employed by ethnologists.

But a GIS is more than an efficient form of cartography. The conventional assumption that it is a tool for quantitative spatial analysis remains one reason why it is often not considered in research that is largely qualitative. However, recent developments in digital technology mean that GIS can deal with many types of qualitative information and as a result 'the use of GIS ... does not necessarily preclude the use of rich and contextualized primary data' (Kwan, 2002, p. 272). Like most ethnological research projects, 'Room to Rhyme' has involved the collection of diverse forms of qualitative data. For example, extensive archival research on mumming has been carried out in the Department of Irish Folklore and the Ulster Folk and Transport Museum. The project has also used a questionnaire survey to gather reliable local information about past and present-day manifestations of the tradition. In addition, many older mummers were interviewed in an effort to fully document the scope of the tradition. Numerous photographs and video recordings of active mumming groups have also been collected. It is possible to integrate all this material into a GIS.

In summary, there are four key reasons why 'Room to Rhyme' provides an ideal case study through which to consider the ways that GIS might contribute to a more sophisticated understanding of a folk practice than could be obtained from conventional ethnological research methods alone. First, it is inherently an issue of spatial and temporal analysis which makes the project suitable for the application of a GIS. Secondly, the research requires the production of ethnological maps to analyse the distribution of the Irish mumming tradition and given that GIS was originally developed to obviate the limitations of paper-mapping, the project provides an opportunity to assess the methodology's potential to enhance conventional paper-based ethnological mapping. Thirdly, the project involves the use of qualitative data, and so the project also provides the opportunity to investigate the integration of qualitative data into a GIS. Finally, however, the project also offers an opportunity of assessing the importance of social factors in the implementation of a GIS, given that the methodology has not hitherto been applied to any aspect of the study of Irish folklore and ethnology.

GIS and the 'Room to Rhyme' Project

The GIS

The first consideration when creating the GIS was the selection of software with which to develop the system. The primary objective of creating the GIS was to store, visualize and analyze the data. However, it was established that the system should be created in such a way as to make it accessible to people with different levels of GIS experience. The ArcView package was selected because it is a powerful yet easy to use tool that provides a user-friendly interface for use by people with little or no experience of GIS. The ArcView GIS application for 'Room to Rhyme' was created in four key steps:

- the locations of mumming performances in Ireland were identified;
- locational data in the form of Irish grid coordinates were collected for each mumming location;
- the geographical and attribute data pertaining to each mumming event were recorded in a database;
- the database was transferred to the ESRI ArcView GIS software to permit the mapping and spatial analysis of the mumming events.

Distribution Analysis

One important objective of the 'Room to Rhyme' project was to explore the geographical distribution of mumming in the North of Ireland. All mumming occurrences were mapped and the past and present distributions of the tradition were compared. In addition, marked variations were observed in the nature of mumming throughout the study area. Thus a spatial analysis was carried out of two different aspects of the plays; namely, mumming group names, and characters.

Analysis of data relating to the custom of mumming in the years 1900-1965 reveals that it was prevalent in counties Sligo, Leitrim, Cavan, Monaghan, Meath and Louth as well as the six counties of Northern Ireland, with a notable concentration around the border areas of counties Donegal, Fermanagh, Cavan and Armagh. When the results of recent fieldwork are mapped and contrasted with earlier accounts, a dramatic reduction in the frequency and extent of the custom is evident. With few exceptions, the custom is now restricted to west Ulster, especially east Donegal, west Tyrone and south Fermanagh. It should be added that the custom is remembered by older generations in many other parts of the study area, and this latent folk memory has provided the basis for recent revival of the tradition in places such as south Armagh, east Derry and the Ards Peninsula of county Down.

The analysis of the attribute data revealed distinct distribution patterns with regard to mumming group name conventions. 'Mummers' (or 'Christmas Mummers') is the name most frequently associated with the tradition, and prevails in the west of the study area. The second most popular name, 'Rhymers' (or 'Christmas Rhymers'), is a form exclusive to the north and northeast of the study area. Excepting the Inishowen peninsula of north Donegal, these distinct distribution patterns correspond broadly to 'west' and 'east' of the River Bann. Lesser-used names include 'Hogmanay Men', a localized form recorded in a limited area of mid-Fermanagh, and 'Wren Boys', a name associated with the custom in south Ulster (including counties Cavan, Monaghan, Donegal and south Fermanagh). The appearance of the name 'Wren' is significant as these areas represent an interface between two equivalent types of Yuletide folk drama – Christmas Mummers and Wren Boys (associated chiefly with St. Stephen's Day, 26th December). As a rule, where one form of folk drama is present in Ireland, the other is not found. The application of the name 'Wren Boys' to the Christmas mumming tradition represents a tacit acknowledgement of the existence of the other. In county Leitrim, on the other hand, where both customs are also recorded but where the tradition of 'hunting the wren' on St. Stephen's Day dominates, Wren Boys are known as 'Mummers'. Further evidence of this cultural overlap is evidenced in the insertion of a character called the 'Wren' into the mummer's play in parts of counties Fermanagh, Donegal and Leitrim. Part of the rhyme recited by the character draws partly on the usual Wren Boy rhyme – 'The wren, the wren, The king of all birds, On St. Stephen's Day was caught in the furze…'

Turning to the characters in the mumming play, analysis reveals that a core number of characters are common throughout the study area. The mummer's play is often described as a hero-combat drama, involving as it does a fight between two warriors. The most favoured pairing throughout the study area is King (or Prince) George versus St. Patrick. There is little evidence for division along sectarian lines with regard to which of the two national representatives in question is victorious. Of the two, it is normally St. Patrick who falls and is revived by the quack Doctor, a pattern that is repeated right across the study area. Furthermore, a second and almost equally popular pairing is that of King George and the Turkish Champion, most likely the two combatants to appear in the earliest versions of the play to have reached these shores. A Scottish influence is apparent in the figure of Galotian (or Galoshin), the central character of Scottish hero-combat plays. Galotian appears in a handful of versions of the play recorded in north Donegal and mid-Antrim.

Other figures who appear repeatedly in the play are the 'Doctor', 'Devil Doubt', 'Beelzebub' and 'Oliver Cromwell'. The latter figure is well established throughout the study area. His presence is significant in that he is a well-known historical figure adopted into the play at an early stage. Cromwell's 'bogeyman' status in Irish folklore renders him an ideal candidate for inclusion in the play, but his notoriety in Ireland is not replicated in England where he is remembered in a much more positive light and not as a subject fit for mocking.

Certain mumming characters have only limited distribution. The character 'Green Knight', for instance, who appears in versions of the play in predominantly nationalist districts of west and south Ulster, may represent an expression of nationalist sentiment. He is dressed in combat 'green' and is typically war-like in disposition although his rhyme is short and his role in the play limited. More interesting still is the 'Jack Straw' character. While he features often in the play, his distribution is markedly western and would appear to owe its origin to indigenous Irish mumming traditions – he is entirely absent from English mumming. As the name implies, Jack Straw is dressed in vegetational costume and disguise and is, in essence, a 'foliage' figure who boasts of his sexual prowess.

Evaluation of the GIS Application

The key objective of the evaluation is to determine the ways that GIS might improve upon and/or complement conventional ethnological research methods. As observed above, social factors related to the prevailing methodological norms of an academic discipline can play an important role in determining the success or otherwise of a GIS application. Therefore, it is important first to consider the social context of 'Room to Rhyme' and its influence on the success of the GIS application.

Social Context

User participation in the GIS development process is viewed as a way of designing the system with greater knowledge and understanding, but perhaps more important, it is seen as a means through which to obtain the end user's consent for the new technology (Reeve and Petch, 1999). Two researchers participated in the creation of the GIS application. A GIS expert developed the GIS application with the assistance of the end user, an ethnologist working on the 'Room to Rhyme' project. Although Butler and Fitzgerald (1997) conclude that social issues and the conflict that arise from them can have a negative effect on the overall success of the application of GIS, in this case the ethnologist's attitude meant that there was little conflict concerning the use of GIS as a research tool in 'Room to Rhyme'. From the outset, the ethnologist was fully aware of the limitations of conventional ethnological mapping, and was keen to try out new technologies and methods that might enhance established ethnological research methods. Thus the ethnologist supported using GIS in the 'Room to Rhyme' project. The most challenging issue in the creation of any GIS application lies in defining what has to be constructed (Bestebreutje, 1997). End users have an important role to play because they have to inform the developer of their requirements of the system. Having no prior knowledge of the system, however, the majority of new users cannot enlighten the GIS developer, a dissonance that makes the critical task of defining user requirements especially difficult and time consuming for GIS development. This problem was encountered during the development of the 'Room to Rhyme' GIS. While wishing to use the GIS, but having no prior knowledge of the technology, the ethnologist was unaware of the strengths and limitations of the GIS and, consequently, was unsure as to what the methodology could bring to the project. While he did gain some knowledge and appreciation of the technology, and was able to help define the system requirements for the GIS application, this factor meant that, despite the positive attitudes to the methodology, the full capabilities of the GIS were underused, the ethnologist perceiving it as a mapping system only rather than a means through which diverse qualitative data elements and media files could be incorporated into a database prior to its interrogation through many different dimensions.

GIS-based Ethnological Mapping

In this case, it appears that the GIS-derived maps did not significantly improve upon the paper-based ethnological maps created by Gailey (1972). This was largely due to the simplistic exploratory mapping needs of the project, in part a reflection of the failure of the ethnologist to realize the full significance of GIS, and thus challenge the mapping capabilities of the technology. However, GIS is often misunderstood as merely a mapping device, whereas the strength of the technology is derived from the potential it provides to interrogate, manipulate and query various forms of data via the digital map display.

Indeed, even given the limited success of the 'Room to Rhyme' GIS in creating innovative ethnological maps, the project showed that GIS-based ethnological mapping provided the project with a level of cartographic flexibility that would otherwise have been impossible to achieve with conventional paper maps. The GIS allowed experimentation with different combinations of scales, colours and symbols in order to create the most visually effective maps. In this way, the methodology provides ethnological mapping with ample scope for trial and error in the process of perfecting the map display. Furthermore, Gailey and O Danachair (1976) argue that a major limitation of paper-based ethnological mapping is that data has to be aggregated or removed to prevent visual analysis being impeded by overcrowding of information. But the exclusion of potentially valuable data can make interpretation more difficult. Unlike static paper maps, a GIS is inherently flexible and this makes it possible to add and remove data from the GIS-based ethnological maps as and when required. Further, GIS can be used to filter and extract particular data records so that they can be analysed separately or compared with other data layers. This function proved particularly beneficial in the 'Room to Rhyme' project, permitting the isolation for individual analysis of particular characteristics and spatial patterns in the distribution of the mumming tradition. Thus, GIS permitted a greater depth of analysis, offering new insights into the factors that have shaped the nature of the mumming tradition at different geographic locations in the north of Ireland.

Data Storage and Access

The evaluation also indicated that, despite the limited exploitation of the full power of the methodology, the diverse functionality of the GIS had benefited other non-geographical aspects of the research process. In particular, the use of GIS transformed the nature of the data storage and data access for the study. In previous studies, a paper-based record system would have been used. As a consequence, different data elements would have been stored in different files, making data analysis difficult and tedious. While qualitative data is often considered to be unsuitable for inclusion in a GIS, recent developments in digital technology mean that the methodology can deal with many types of qualitative information. Indeed, the study demonstrated that diverse qualitative data elements and media files could be incorporated into the GIS database. Further, the storage of all data in one database location makes it possible to access, view and reassess diverse forms of data via the computer desktop.

Moreover, data collection for the 'Room to Rhyme' project took place over a period of two years. Conventional paper-based ethnological maps cannot be updated without considerable effort and expense when additional fieldwork and/or additional archival material become available. Conversely, the dynamically linked GIS database and map display make it possible to update and review data easily. It is anticipated that this function would be particularly beneficial to long-term ethnological research projects that require constant updating and reviewing of data. Given that most ethnological research involves qualitative data in diverse forms, the ability of GIS to integrate, store, organize and access data is of notable value to many aspects of the ethnological research methodology.

Further Analysis

The use of GIS in 'Room to Rhyme' has also generated potential new research directions in future studies of the mumming tradition. For example, the GIS could be used to analyse the distribution of the custom within particular regions and localities. In the case of mumming, the distance travelled by individual groups from their home base reflects the true extent of the custom in the local district. While topography and ease of communication heavily influence this hinterland, previous studies using conventional paper maps have excluded such information to prevent overcrowding of information interfering with visual analysis. As an alternative, previous studies have used a simplistic and crude approach that places a circle around the home base to represent the extent of custom in the localized area. With the GIS-based approach, it is possible to incorporate additional data about topography and communication into the map so that the true extent and density of the mumming custom in localized areas could be more accurately defined. Given that the majority of ethnological studies tend to focus on the cultural traits of localized communities (Dow, 1994), the ability of GIS methodology to provide accurate high-resolution ethnological maps is particularly beneficial to this type of research question. Moreover, GIS is especially useful in ethnological studies within an Ulster context because it can map the distribution of a cultural trait within a locality with a resolution that matches the underlying finely differentiated cultural mosaic of the province.

Regardless of the relatively straightforward nature of the 'Room to Rhyme' study, the involvement of GIS has proved advantageous in a number of ways. Thus, the final mapping product alone should not be used to gauge the success of a GIS application. Therefore, other ethnological case studies requiring more complex forms of spatial and temporal analysis and a full digital integration of archival and aural evidence would certainly benefit from the intervention of a GIS.

Conclusions

While previous studies have established a reasonably definitive statement of the broad geography of the Irish mumming tradition, it has been necessary to document the tradition and explore the variations in the nature of the mumming at different locations in the North of Ireland. 'Room to Rhyme' sought to address this need through the creation of a GIS that recorded a variety of data and permitted GIS spatial analysis of different aspects of the tradition. This provided detailed information about the geographical variations of aspects of the mumming play, and as such generated new insights and interpretations of the factors that have shaped the nature of the tradition at different locations.

Although GIS is still used relatively rarely in ethnological research (or in heritage studies more generally), its incorporation into 'Room to Rhyme' suggests that there is a real potential for GIS as a research tool in the domain of cultural heritage. GIS can enhance research methodologies in these areas in three principal ways: by eliminating the problems associated with production of paper maps; by assisting with other non-geographical aspects of the research process such as data storage and management; and by generating new types of information that provide greater insights and a richer understanding of cultural elements. These points echo the conclusions of the case studies by Dow (1994) and Ebaugh et al. (2000) discussed earlier.

However, social issues and personal attitudes play an important role in determining the incorporation, acceptance and overall success of GIS applications. Given that the vast majority of cultural researchers generally and even those in more spatially aware heritage studies have so far been resistant to the use of GIS technology, its promotion requires a change in attitude towards the use of the methodology as a cultural research tool. While, the capabilities of the GIS were by no means fully challenged by its application to 'Room to Rhyme', largely because social factors meant that the full import of GIS was not understood by the ethnologist, this case study still illustrates how GIS can enhance and inform the study of cultural heritage through time and across space.

Acknowledgements

I am grateful to the staff of the Department of Irish Folklore, University College, Dublin.

References

Bestebreurtje, J.G.A. (1997), '*GIS Project Management*', M.Sc. thesis, Department of Environmental and Geographical Sciences, Manchester Metropolitan University.
Butler, T. and Fitzgerald, B. (1997), 'A Case Study of User Participation in the Information Systems Development Process', in E.R. McClean and R.J. Welke (eds), *Proceedings*

of the 18*th* *International Conference on Information Systems*, ICIS, Atlanta, pp. 411-26.

Campbell, H. and Masser, I. (1995), *GIS and Organizations: How Effective are GIS in Practice?*, Taylor and Francis, London.

Dow, J.W. (1994), 'Anthropology: The Mapping of Cultural Traits from Field Data', *Social Science Computer Review*, **12**, pp. 479-92.

Ebaugh, R.H., O'Brien, J. and Chaftez, J.S. (2000), 'The Social Ecology of Residential Patterns and Membership in Immigrant Churches', *Journal for the Scientific Study of Religion*, **39**, pp. 107-16.

Gailey, A. (1972), 'Towards an Irish Ethnological Atlas', *Ulster Folklife*, **18**, pp. 121-25.

Gailey, A. (1978), 'Mummers' and Christmas Rhymers' Plays in Ireland: The Problem of Distribution', *Ulster Folklife*, **24**, pp. 59-64.

Gailey, A. and O Danachair, C. (1976), 'Ethnological Mapping in Ireland', *Ethnologia Europaea*, **9**, pp. 14-34.

Kwan, M.P. (2002), 'Is GIS for Women? Reflections on the Critical Discourse in the 1990s', *Gender, Place and Culture*, **9**, pp. 271-79.

McLafferty, S.L. (2002), 'Mapping Women's Worlds: Knowledge, Power and the Bounds of GIS', *Gender, Place and Culture*, **9**, pp. 263-69.

Murayama, Y. (2001), Geography with GIS', *Geojournal*, **52**, pp. 165-71.

Philip, L.J. (1998), 'Combining Quantitative and Qualitative Approaches to Social Research in Human Geography – An Impossible Mixture?' Environment and Planning A, **30**, pp. 261-76.

Pavlovskaya, M.E. (2002), 'Mapping Urban Change and Changing GIS: Other Views of Economic Restructuring', *Gender, Place and Culture*, **9**, pp. 281-89.

Petch, J. and Reeve, D. (1999), *GIS Organizations and People: A Socio-technical Approach*, Taylor and Francis, London.

Schuurman, N. and Pratt, G. (2002), 'Care of the Subject: Feminism and Critiques of GIS', *Gender, Place and Culture*, **9**, pp. 291-99.

THEME II:
THE PUBLIC/OFFICIAL CREATION
OF PLACE IDENTITIES

Introduction to Theme Two

The Editors

The main concern of Theme One was the nature of the relationships of places and identities: in this second set of chapters, the debate is broadened to address such key questions as: 'who deliberately creates place identities and why do they do that?' The focus here is upon the conscious use by public jurisdictions and official agencies, implicated in the commodification of senses of the past, to reinforce or even create an identification of people with specified places. Inevitably, because of two related dichotomies inherent in all questions of heritage and identity, tensions and conflicts emanate from these processes. The first is between the collective and the individual: the second between the official and the unofficial.

The paradox is that most of the attention paid by policy-makers and academic observers to heritage, and much cultural expression and performance in general, focuses upon the collective and the official. This occurs to such an extent that the practice of heritage could easily be viewed as effectively an official monopoly regulated by government legislation, subsidy, and policy, which is managed by official public agencies. In 1862, the-then Dutch Prime Minister was able to state that 'art is not the business of government: the government is not a judge of science or art', which should be properly left to individual selection, taste, enjoyment and finance. It took only a century to arrive at the situation in which the state and its agencies at many levels and scales appear to exercize an almost unchallenged control over art as expressed through the concept of heritage. Arguably, however, an individual's sense of the past derives most strongly from the mundane minutiae of personal and family life experiences, over which official agencies and their policies have no direct cognizance or influence. Individual memory, therefore, may be more important than any postulated collective memory in determining the self-identification of the individual with the group or with a sense of place.

Equally there are unofficial senses of time and place that may exist independently from the official forms. For example, a visitor to a museum, or an observer of a monument, may interpret their meanings in a form that is totally dissonant to that intentionally projected by authority. More important, however, heritage is also a mode of resistance and its unofficial forms can often be explicit rejections of statist ideology and the agencies charged with its implementation.

Public bodies at various levels in the jurisdictional hierarchy may foster a sense of place at various spatial scales ranging from the local through the regional to the national, supranational and global. It is increasingly the case that heritage, so long the domain of the national, is now being defined against that scale of belonging. In part, this reflects the wider processes of globalization and their ramifications for the increasing importance of regional and local scales of human activity at the expense of the national. The elision of the national in the four studies addressing Theme Two is not, however, a deliberate refutation of its importance. On the contrary, the link between the nation state, and especially the deliberate creation of the mystical entity 'the nation', dominates much of the literature dealing with heritage, culture and identity. Thus while none of the cases discussed here addresses the national scale directly, the 'national' remains omnipresent in one guise or another, providing the comparative context for the studies of the manifestations of heritage at local, regional or international scales that follow.

Any simple idea that heritage nests in a hierarchy of spatial scales is dispelled in Chapter 7 in which the confusion and conflation of local, regional, national and, in this case, also trans-national scales of identification are complete. As Miller shows, the links between regiments and places in the British military recruiting system are intimate, enduring and bilateral. Local place identification contributes *espirit de corps* to the regiment, which, in turn, together with its museums, ceremonies and even the physical presence of its members, can contribute to a local sense of place. There is, however, a distinction between political and administrative jurisdictions that inevitably operate within clearly demarcated hierarchies of spatial scales and popular place identifications, which frequently do not. The clearest illustration of this is to be found in the recruitment of people from one national jurisdiction into the army of another, while simultaneously retaining and exploiting the local sense of place. In Miller's study, 'Irishness' makes a valued contribution to a regiment's sense of group identity, while Irish places are slowly re-establishing identification with 'their' regiments. Both sets of processes events transcend the purely national attribution of identity.

Van Dam's account of the new Canadian administrative territory of Nunavut [8] makes three significant contributions to this discussion. First, it challenges the simple dichotomy between a top-down imposed and a bottom-up organic identity. Is Nunavut a creation of a Canadian federal government reacting to its own concerns and fulfilling its own vision of the culture and aspirations of an indigenous population, or is it the logical resultant of a local desire to express a distinctive sense of place through a jurisdictional entity? Is Nunavut little more than a national theme park, preserving and presenting a fossilized idea of Inuit or the precursor and model of self-determination and self-expression for many such indigenous groups in settler societies elsewhere? Secondly, it is by no means clear whether the jurisdictional scale of Nunavut or Canada represents the concept of 'nationhood' in a nineteenth-century European sense. Nunavut is more than a territorial convenience for administering services at the sub-national scale: it is a proto-province intended to reflect and express the Inuit as nation. Thirdly, Van Dam raises a wider question about the relationship of a certain type of culture and economy and the identification of people with places. Does a society whose culture derives from a, now largely superseded, hunter-gatherer economy have a sense of place that is significantly different from one based on a traditionally more settled way of life?

The commemoration through monuments, place names and ceremonies of those who died in the service of the state evokes powerful individual and collective emotions, which the state then memorializes for various purposes. These issues are particularly complicated in divided societies such as Northern Ireland where such memorials are touchstones of both individual loss and collective political allegiance. In Chapter 9, Switzer examines the changing roles of war commemoration in Northern Ireland beginning with the Great War and culminating in the more recent 'Troubles'. As she argues, although the outward forms of commemoration have not changed greatly since the 1920s, the meaning of the rituals has been broadened and the memorials rededicated to include a wider range of conflicts. Specifically, public commemoration of members of the security forces killed during the Troubles draws on elements of world war commemoration, and raises important issues about the purposes to which commemoration is put. Again, this demonstrates the conflation of official and unofficial and the interaction of the national with the local in heritage. Given the absence of an agreed national identification amongst Protestants in Northern Ireland and a 'national' commemoration of the victims of the Troubles, security force memorials serve to claim existing war memorials for the Protestant community. They can also be seen as providing a focus for the public expression of grief, although graves more powerfully serve this purpose for the relatives of security force personnel.

The final case [10] examines the history of the designation of world heritage sites in Mexico. The identification of the individual with the earth as common home planet or with humanity as a common social group is no novelty but has been intrinsic to global universalizing religions and ideologies. What is new, however, is the attempt to operationalize such ideas through agencies, practices and designations responsible for selecting, protecting and managing what is regarded as the heritage of all mankind. The idea of the existence of a world heritage is, on the one hand, so obviously self-evident that almost everyone, almost everywhere, regards the Taj Mahal, the Great Barrier Reef or the 'Mona Lisa' as in some way their own tangible heritage. This identification is expressed through their government agencies, tourism visits and above all popular concern. On the other hand, however, this global identification is only weakly matched by any international operational concept of ownership, financial support or management responsibility. The ambivalence of UNESCO and its agencies in regard to national claims can be uncomfortable and on occasion disastrous. Van der Aa raises not only the question of whether world heritage sites in Mexico are 'ours' or 'theirs' but also whose Mexico – Spanish, Indian or Mestizo – is being expressed through the changes in the pattern of selection over the years. The case has clear elements of post-colonial nation building, of shaping identity in settler societies and interethnic competition and conflict, each of which could be echoed elsewhere.

Chapter 7

Irish Regimental Heritage: Representations of Identity and War in a Climate of Change

Kenneth J.S. Miller

Introduction

'Heritage is history processed through mythology, nationalism, local pride, romantic ideas or just plain marketing, into a commodity' (Schouten, 1995, p. 21). The significant role that heritage can play in the definition of identity raises questions about the extent to which Irish regimental heritage contributes to the cultural identity of Ireland, and to what extent the appropriation of Irish identity in the British Army has had an effect on defining 'Irishness'. The Irish military experience provides a rich and interesting basis for considering the significance of heritage and in the sense of identity. The military heritage of Ireland shows much contrast between the ways in which different groupings of the Irish people remember the past, interpret the past and use it to create their own senses of place and time.

The Irish Defence Force has not yet developed much of a heritage. Since the creation of the Irish Republic (a neutral state) its operations have been limited to United Nations operations and home defence. The British Army now contains just two explicitly 'Irish' regiments: the Irish Guards and the Royal Irish Regiment and two cavalry regiments maintain a sense of Irish identity within amalgamated structures (Hallows, 1991; Harris, 1999). However, these contemporary parameters of Irish military identity provide little indication of the rich historical record that Irish soldiers have achieved over the centuries, nor of the different manifestations of this heritage. The close relationship that was developed between British regiments and their specific regional recruitment territories (within the British Army's regimental system) has created, historically, a conscious and explicit mutuality between urban or rural areas and 'their own' regiments. Arguably, this pattern was equally significant within the context of the Irish regiments although the recruitment areas tended to be much bigger and culturally more varied than those in Great

Britain were.

 Significantly, something of a change in the perception of military heritage is evident, both in Great Britain and in Ireland, raising questions about the way it is, and *should* be, represented (Miller, 1998; 2001), and the potential for its changing use, relative to transformations in the recognition and value of this heritage.

The Pluralism of Irish Identity

'There have been and there are many Irelands' (Smyth, 1997, p. 19). The pluralism of Irish identity is probably the most recognisable and defining factor that has shaped its past and that created different manifestations of heritage. The course of history was defined by such pluralism, and yet the *shaping* of that past is a contemporary process – one that changes in response to diverse influences.

> It is through literature and its readings, as well as its geography, that Irish place is defined and redefined, constantly negotiated as society is contested along its many and varied axes of differentiation by its myriad actors and their conflicting motivations (Duffy, 1997, p. 81).

'Irishness' is a highly differentiated concept due to cultural divisions and the apparent rigidity of interpretation by different groups within Irish society. 'Irishness' means so many things to different people, not only within Ireland itself but also through the appropriation of 'Irishness' in many parts of the world in different ways and to different purposes. This has created conflict and confusion in how Irish identity can be defined, probably more than in most other national identities.

 The contemporary shaping of the past and the understanding of the past is a selective process used to 'legitimate and validate...present attitudes' while '...any social reality must be referred to the space, place or region within which it exists. Places are invented, a myth of territory being basic to the construction and legitimisation of identity...' (Graham, 1997a, p. xi). This 'myth of territory' is heightened by the expatriate and other exported forms of 'Irishness' which, in turn, generates a new expectation and pressure on the culture. Graham (1997b, p. 7), writing about the west of Ireland, described how it became '...an idealised landscape, populated by an idealised people who invoked the representative, exclusive essence of the nation through their Otherness from Britain.' Within a broader context, it can be reasoned that this defines a mystical idealism that was created for Ireland as a whole (Duffy, 1997; Mitchell, 2000) – an emerald isle – and formed the basis for Britain's consciousness of Irish identity for generations. The internal divisions of Irish identity has not precluded the possibility of the creation of an idealized Ireland – the creation of a cultural landscape, for or by external consciousness or consumption, that overarches any conflict or confusion about what it is to be Irish.

 Historically, 'Irishness' has been seen as an identity that is distinct from other British military identities. The Irish soldier became a romanticized symbol of courage, loyalty, humour and sacrifice (thanks to the way he has been portrayed by Kipling and others of the same genre), but to the Irish Republic such perspectives are overshadowed by the colonial implications of that military service. Certainly, this favoured image is not exclusively the claim of the Irish, as this has also been held true of the Scots, the Welsh and a number of other identities. The concept of identity, grounded in the identification of similarities between 'owners' of a shared identity and their difference from others, is reflected in discussions on how the imagery of the kilted Highland soldier helped to define the identity of Scotland (Wood, 1993; McCrone et al., 1995; Trevor-Roper, 1995). Connelly (1997) compares the appropriation of the Highlander image with the apparent lack of an identifiable equivalent in Ireland. Although such distinctiveness is, perhaps, less readily identifiable in 'Irishness', there is a rich popular symbolism and a manufactured cultural imagery of Ireland that has been mobilized by the military as a means of highlighting difference and promoting hegemony, which is equally identifiable in civilian contexts.

Many non-Irish people in the British military context have adopted Irish identity. In World War I, the men of the Channel Islands were allowed to choose the units in which they would serve and almost without exception chose to serve in Irish regiments, although any specific reason for this is difficult to determine. The Irish community in London formed the London Irish Rifles in 1859. By 1861 the battalion had a strength of 1,200 and was thus the third largest unit in the British Army. There was a strict rule that only Irishmen could be admitted; thus the unit maintained a unique Irish cultural entity 'in exile' and actively protected its Irish 'pedigree'. In WW1, its two battalions saw active service in the London Divisions (47th and 60th) and not in Irish ones (Harris, 1999). By then, it is evident that despite its Irish image, it was a *London* regiment and not an Irish one.

The London Irish Rifles – now D Company of The London Regiment (Territorial Army) – has now only around 10 per cent genuine Irish recruits. 'A soldier wearing the London Irish cap badge is more likely to have roots in the West Indies than in the west of Ireland.' 'Irishness' has become little more than a brand image and the identity a transferable commodity that rests more on the fame of the regiment's achievements than on its cultural origins. Here the reality of the identity is maintained in a different time and space. The redefinition of self is nothing new in the context of Irish history (Connolly, 1997, pp. 50-1) as Scottish settlers in Ulster very quickly assumed an 'Irish' identity – albeit distinct from the Catholic Irish identity. 'Identity is not "given" in human nature, but is learned through social interaction and communication in a complex of social structures, set in specific and distinctive places and epochs.' (Douglas, 1997, p. 155)

At the outbreak of World War I, an enthusiastic and patriotic rush to 'join the colours' was, for most, a simple and obvious procedure of joining the local regiment. However, leaders of the Irish Catholic immigrant population on Tyneside determined they should have their own Irish regiment, instead of simply joining the Northumberland Fusiliers or the Durham Light Infantry (Sheen, 1998). Within months, four battalions had been raised, although not all the recruits were Irish or of Irish descent. Although Irish in identity (they formed the Tyneside Irish Brigade and wore a shamrock on their sleeve and as a cap badge backing), these units remained part of the Northumberland Fusiliers, wore their cap badge, and served throughout the war in non-Irish Divisions. This is perhaps an example of what Graham, et al., (2000, p. 19) describe as '…a people cut off from their past through migration…[wishing to]…recreate it, or even "recreate" what could or should have been but never actually was.' Certainly, their self-identity was Irish and not Tyneside. The perceived need for a distinct Irish identity among the Tyneside recruits demonstrates the strength of an expatriate cultural identity. It has also been suggested that illiteracy among immigrant Irish was as much a reason for the emergence of both the Tyneside and London Irish units. Where this could have been a problem in the rest of the army, it was not seen as an issue among the Irish units.

This move is understandable given the fact that the Irish on Tyneside were marginalized people and did not have a ready identification with the rest of the community (Sheen, 1998). A similar idea was attempted on Merseyside but was rejected as the 8th (Territorial) Battalion King's (Liverpool) Regiment was an Irish unit and there was no desire for more. (Indeed, many Irish from this area, from the London Irish and from Ireland too, moved to Tyneside to join the Irish Brigade.) However, large concentrations of Irish also existed in Manchester and in Glasgow and yet there is no evidence that there was ever a perceived necessity to create Irish battalions of The Manchester Regiment, the Highland Light Infantry or the Royal Scots Fusiliers. The Irish from these areas merely joined their local regiment or went home to join Irish ones. Douglas (1997) describes how, 'An identity is expressed and experienced through communal membership, awareness will develop of the Other – identities and groups with competing and often conflicting beliefs, values and aspirations. Recognition of Otherness will help reinforce self-identity, but may also lead to distrust, avoidance, exclusion and distancing from groups so defined' (Douglas, 1997, p. 152). This idea of defining one's own identity through distancing oneself from 'Otherness' sits at the heart of Irish cultural divisions and, indeed, at the heart of the organization of military units.

The Irish and the British Regimental System

The regimental system is a distinctive feature of British soldiering. The system gains its reputed strength from having geographically defined recruitment areas for each regiment and by maintaining a sense of regional identity as the basis for the development of *'esprit de corps'* (Beckett, 1999). This approach towards recruitment initially related only to the first 25 infantry regiments of the line and developed as a feature of other regiments over time, until in 1881 when specific directives established it for all infantry regiments. The extension of the approach to the cavalry was officially initiated in the 1920s. With respect to the Irish units, the earliest example was the 18th Foot, raised in 1683, and recruited from throughout Ireland, though mainly from the south, and very soon became known as The Royal Regiment of Ireland. This geographical relationship remained instrumental to the identity and success of the regiment throughout more than 200 years of service.

To this day, the regimental system remains (although the rigor of the concept must now be questioned in the light of defence cuts, amalgamations and the pressure to meet recruitment targets) and both the Irish Guards and the Royal Irish Regiment (27th [Inniskilling], 83rd and 87th) continue to recruit from *all* Ireland. The cultural divisions in Irish society seldom influenced the recruitment of Irish into British Army regiments. The Royal Irish Regiment's designated recruitment area reaped mostly Catholics, but the Royal Irish Fusiliers and the Royal Irish Rifles recruited in the north of Ireland across the cultural divide (Harris, 1999). This ability to circumvent or overcome societal divisions within regimental structures is also evident in the context of British India. Many sepoy regiments freely recruited Sikhs, Muslims and Hindus, both in the days before the Mutiny and after it. Officered mostly by British, the common differences between the societies in India were, to all intents and purposes, insignificant. '...the crucial factor holding the army together in the face of death at the battlefront is not patriotism...but *esprit de corps*' (Jeffrey, 2000, p. 97). The importance of *esprit de corps* in battle is easily understood, but most regiments spend proportionately little time in combat and are usually engaged in very routine, even mundane and tedious duties. For any system to be able to create a collective sense of identity which surmounts such fundamental differences as those within Irish (or Indian) society suggests that the creation of regimental identity with a unifying culture is a very powerful process. In the context of regimental identity the sense of belonging to place is very strong. It would seem that the common recognition of being Irish was strong enough to overcome any differences in religion and culture. The closest any of the Irish regiments came to any problem with the cultural divide was in 1921 when the name of the Royal Irish Rifles was changed to the Royal Ulster Rifles. The change was never understood by the regiment and was not popular, having been imposed on them very suddenly 'from above'.

Regimental Culture and Identity

For most regiments, the shared geographical (and thus common cultural) origins of the recruits provide the basis for the nurturing of the regiment as a cultural entity. This is built on and stimulated by specific regimental experiences and differences. 'The very ubiquity of the relationship between politico-cultural institutions and territoriality suggests that a representation of place is a key component in communal identity, whatever the scale' (Graham, 1997b, p. 6). It is the operationalization of territoriality that lies at the heart of regimental culture.

The different aspects of culture detailed by Hofstede (2001), values, rituals, heroes, symbols and practices, provide a basic framework for illustrating how these are operationalized in the context of a regiment, but are not enough in fully understanding the dynamics which influence military heritage or its value in society. Any military unit can point to most of these cultural elements, but it is the fundamental underpinning of all this to place (and thus a stronger sense of identity) which makes the British example particularly strong. Although Douglas (1997, p. 152) suggests that identity is not immediately recognisable, the way that it is presented in the regimental context *is* to create immediate and unambiguous recognition. Murray (2001) expresses the role of regimental music in carrying an audible signal over distance to friend or foe that a particular regiment (and unmistakably that one) was advancing on the position. The dress, symbols and other elements re-affirmed this signal at closer distance: 'culture as signification' (Graham, et al., 2000, p. 23). In the Irish context, the 'regional diversity and cultural heterogeneity' (Graham, 1997b, p. 10) and 'strong regional sub-cultures' discussed by Smyth (1997, pp. 19-21) summarise precisely the richness of diversity that was appropriated by the British regimental system in the creation of the Irish regiments. It is such differences that were used to build the regimental identities and the *esprit de corps* that, in part, made the Irish so successful as soldiers.

The Representation of Irish Identity in the British Army

There is much evidence of heritage within the British military establishment to legitimize Ireland and Irishness as being British. Graham (1997b) points to the need to revise the representations of Ireland's past and its imagery of place, as traditional renditions of Irishness now have little relevance. 'Irishness', as expressed in the regiments of the British Army, relates very much to this traditional imagery. "Invented tradition' is taken to mean a set of practices, normally governed by overtly or tacitly accepted rules and of a ritual or symbolic nature, which seek to inculcate certain values and norms of behaviour by repetition, which automatically implies continuity with the past' (Hobsbawm, 1983, p. 1). Thus, the use of traditional symbols in the creation of regimental identities extends the claimed legitimacy of the regiment's Irish 'genes'. This is perhaps most apparent with the Irish Guards. Raised in 1900, the regiment defined itself through the traditional imagery and symbolism of Irish regiments and those of the older Guards regiments. These symbols have created a sense of identity for the regiment that make it *seem* far older than a century, '...by conveying an idea of timeless values and unbroken lineages' (Graham, et al., 2000, p. 19).

The need to establish such a lineage not only serves the internal identity needs of the regiment but also that of the continued appropriation of Irish identity in British nationhood. However, it is an imagery of Ireland which is arguably more British than Irish in origin. Similar to the appropriation of Highland (and, much later, Lowland) identity by the British Army, selective approval of forms of dress, music, symbols and colours have helped to define not only those regimental identities but have influenced the reformation of the originating identities (as defined by territorial recruitment areas). The legitimization of the British Army's interpretation of Highland identity has had a profound effect on establishing what is now accepted as *Scottish* identity in non-military contexts. The British Army's use of Irishness as part of its own heritage is reinforced by the contemporary view that whatever British Irish regiments may do to express their identity, 'the soldiers of the Irish Defence Forces do not engage in all that as they do not have to try to be Irish'. Notwithstanding this different view towards tradition in the Irish Republic, the leaders of the Irish Free State also recognized that the creation of such identities was important. 'The fostering of a distinctive cultural identity was duly enshrined as a central objective of the new state that was created after the upheavals of 1916-23' (Connolly, 1997, p. 43).

The Harp surmounted by the Crown is the most recurring symbol used by the Irish regiments in their cap badges and in their regimental insignia. Shamrocks recur, both in natural and stylized forms. (Queen Victoria sanctioned the wearing of the Shamrock on St. Patrick's Day – a tradition that continues to this day.) The image of Enniskillen Castle was used as the symbol of the 6th Inniskilling Dragoons and the Royal Inniskilling Fusiliers and was worn in their cap badges, collar dogs and on the drum major's baton. This graphical demonstration of their links with the town was reciprocated with stone images of a Fusilier and a Trooper being displayed on the clock tower of Enniskillen Town Hall. Jeffrey (2000) explains how at the start of World War One there was significant debate about the creation of distinctively Irish divisional badges, illustrating the perceived importance of being represented in the right way. Colours of Hunting Green (originally used because many of the Irish regiments were Rifle regiments) – has become more associated with Irishness. St. Patrick's blue was another innovation of Queen Victoria and is worn in the plumes of the Irish Guards' bearskin caps. The wearing of the 'Caubeen'– an ancient Irish head-dress, and saffron kilts by Irish pipers are innovations that have gradually been adopted as being standard symbols of Irish soldiers, as has the tradition of the Irish Wolfhound regimental mascots of the Irish Guards, the Royal Irish Regiment and the London Irish Rifles.

Music has long been an important feature of military tradition and cultural identity, and has been used, '...to excite cheerfulness and alacrity in the soldier' (Murray, 2001, p. 1). 'The concept of the Regimental March, that is a marching tune associated with one particular regiment...[is] a potent means of creating that sense of exclusive identity...the great majority of British regiments marched past to folk tunes associated with their home county or area' (Murray, 2001, p. 172). For Irish regiments, the most favoured tunes have been *St. Patrick's Day, Minstrel Boy, Killaloe, Tipperary, Garry Owen* and *The Wearing of the Green*.

Celtic crosses seldom appear in regimental insignia, although the memorial to the 16th (Irish) Division in Flanders is a Celtic cross. Use of the St. Patrick's Cross corresponds with traditions among some English and Scottish regiments to include religious symbols in their badges. Much of the symbolism of Ireland has, over the years, been politicized and even appropriated by paramilitary groups. This is something that the Irish regiments have been sensitive to and have tried to avoid as much as possible.

Connolly (1997, p. 44) discussed the, '…inescapable connection between the popularity of Irish history and its controversial character,' and issues of how the interpretation of this is controlled and by whom. The 'ownership' of 'Irishness' seems less disputed than the definition of Irish identity. The cultural divisions in the pluralism of 'Irishness' focus more on religion, class, and politics. Geography seems to be an important factor within Ireland itself but not so much on an international stage. Although the purity of Irish regiments during WW1 was seen to be important to some in Ireland, and the need for this fuelled a push to enlist (Jeffrey, 2000, p. 22), the longevity of this was relatively short lived and there is little evidence to suggest that this remained an important consideration. The 10th (Irish) Division was seen to be the 'purest' of the three divisions raised in Ireland – with around 70 per cent of the men and 90 per cent of the officers being Irish (Jeffrey, 2000, p. 41). However, by 1916, the number of available replacements from home dried up and the division became 'progressively less 'Irish'…' (ibid., p. 59).

The Irish in War and Remembrance

The regular British Army, at its height, comprised four Irish cavalry regiments and nine regular infantry regiments. However, the Irish have served under many flags as mercenaries (the 'Wild Geese') driven far afield by economic hardship and a lack of alternatives. Very many Irishmen served in India in the army of the Honourable East India Company prior to it being subsumed into the British Army following the Great Indian Mutiny of 1857, again largely as economic migrants (Bredin, 1987). During the Napoleonic Wars the need for recruits was high and Ireland provided so many of the recruits then and in the following decades that in 1830 it has been estimated that around 42 per cent of soldiers in the British Army were Irish (Bredin, 1987; Beckett, 1999). Ireland, also, has produced some of Britain's most celebrated military figures: Wellington, Roberts, Wolseley, Kitchener, Alexander, and Montgomery (Beckett, 1999).

In World War One somewhere between 200,000 and 250,000 (Jeffrey, 2000) Irish *volunteered* (conscription was never extended to Ireland) to fight for the British Empire – many of these serving in the 10th (Irish) Division, the 16th (Irish) Division or the 36th (Ulster) Division. 50,000 of them died. Although the number of Irish volunteers is impressive, the number should not be taken as a direct indication of patriotism. 'For those who "took the King's shilling", it is argued that the money [was] the important component of the contract, not whose it [was]' (Jeffrey, 2000: p. 18). Also, it is evident that there was no apparent imbalance between volunteers from the catholic and protestant communities, though volunteers were more likely to come from the industrialised areas of Belfast and Dublin.

During World War Two many Irish volunteered for active service in the British and other allied forces. In addition to the 38th (Irish) Brigade (a specific request of Winston Churchill, recognizing the fighting potential of a wholly Irish Brigade there were also the Irish Fusiliers of Canada and the South African Irish Regiment. To this day, a great many men from the Irish Republic continue to enlist in the British Army, comprising as much as 20 per cent of the Royal Irish Regiment and somewhere between 10 per cent and 15 per cent of the Irish Guards.

The heritage of World War One experiences reflects many common aspects throughout the countries that fought and suffered for the British Empire. The memorials and cemeteries that rose up on the battlefields took on iconic significance (Winter, 1995). Notwithstanding the general nature of this, to the people of Canada, Australia, and, to a lesser extent, South Africa and New Zealand, the battlefields of WW1 were redefined as the spiritual birthplaces of true nationhood (Graham, et al., 2000). In the struggle for Vimy Ridge in 1917, the unity of Canada was confirmed. In the disaster of the Gallipoli Campaign (in which the 10th (Irish) Division fought), Australia and New Zealand found maturity. At Delville Wood the South Africans affirmed their identity. Although such formative heritage has been claimed by these nations, a similar heritage for the Irish has been stifled through the cultural and political upheavals of 1916-1922, although the loyalists have been successful in claiming the military heritage of the Irish for Ulster (Graham, 1997b): '…the attempt by a powerful social group to determine the limits of meaning for everyone else by universalizing its own cultural truths' (Graham, et al. 2000, p. 24). The Somme has become a resonant high point in the legitimization of Ulster's British identity. Whereas the Republic of Ireland allowed their experiences and achievements of WW1 to be put aside, the regimental museums, war memorials and the Somme Heritage Centre have helped sustain the collective memory and politically inspired claim of the loyalty and Britishness of the North. It may be true that attitudes in the Republic are changing but it is only in very recent years that this heritage is being explored and that any sense of a collective achievement of the Irish is being suggested.

On 1st July 1916, the first day of the Battle of the Somme, the 36th (Ulster) Division suffered terrible casualties. Although the moral achievement of this sacrifice has been appropriated by loyalist Ulster (Graham, 1997c), arguably it was the Irish Catholics who made the greatest sacrifice. 2208 men of the Royal Inniskilling Fusiliers were killed – the greatest loss of any regiment on any one day in the history of the British Army (Middlebrook, 1994). It was the Catholic 10th Battalion (The Derrys) who sustained the greatest proportion of this loss.

'The particularly concentrated nature of the Ulster Division, not just socially but also in terms of its religion and politics, meant that its losses on the first day of the Somme, grievous enough in themselves, had a disproportionately great impact back home' (Jeffrey, 2000, p. 57). By the end of 2nd July, the Division had lost 5,500 killed, wounded or missing.

The crowning achievement of the WW1 Irish soldiers was in 1917 at Messines Ridge. Here, the 16th and 36th Divisions advanced together against a formidable German front, and won the day (Gilbert, 1994). 'It was the "first completely successful single operation on the British front since the outbreak of the war", and the casualties were 'incredibly light', a mere 1,400 killed and wounded for both divisions' (Jeffrey, 2000, p. 61). The Irish achievement at Messines was soon reversed when the two divisions fought together a second time, at Langemark in August 1917, and were 'broken to bits' in a heavy defeat (ibid., p. 139). However, a number of the leading nationalists of the day saw Messines as an example of how the 'two Irelands' could in fact put aside their differences and unite. There was a genuine hope that the joint experience of the war would help to settle the differences in Irish society (ibid., p. 63). Such experiences were enough for Australians and Canadians to create a new identity, and yet at the time this was no watershed for the reconciliation of the Irish. Nor can it be said that the events of the Easter Rising the year before automatically overshadowed the achievements of the Irish Divisions. The event merely passed into the 'National Amnesia'.

A memorial to the 36th (Ulster) Division was erected at Thiepval, and other memorials were erected for the 10th (Irish) Division in Salonika, and 16th (Irish) Division in Wytschaete, Flanders and Guillemont in France: granite crosses inscribed with '*Do chum Gloire De agus Onora na hEireann*' (To the Glory of God and Honour of Ireland). Gallipoli is the only major World War I location without a particular memorial to the Irish who fought there (Jeffrey, 2000) – demonstrating how the well-developed heritage of the ANZACs has overshadowed the contributions made by other Imperial soldiers, let alone the soldiering of the Turks, claiming the Gallipoli heritage as Australian.

Even following the war, official acknowledgement and remembrance of the Irish losses was slow to emerge, with the creation of the National War Memorial Trust (1919) and Ireland's Memorial Records (1923) and the eventual dedication of the Irish National War Memorial in 1938. The presence and significance of war memorials differs greatly between the north and south of Ireland, as was evident by the neglect the INWM suffered. It was finally restored in 1998.

Increasing interest in the heritage of Irish participation in WW1 is reflected in the 1998 opening of the 'Island of Ireland Peace Park' at Messines, Belgium. An Irish Round Tower stands as a symbol to promote peace in Ireland by commemorating all those who lost their lives during the First World War. It is located in the position where the 16th and the 36th Divisions fought side by side in June 1917 (Jeffrey, 2000). Connolly (1997, p. 60) points out the need for a specific process of invention and reinterpretation. '...it could be argued that such invention is an essential part of the creation of any workable political community.' Just as nationalists in time revaluated the achievement of the Easter Rising, new perspectives have emerged with regards to other wartime experiences. The Canadian experience was merely a limited military success achieved at immense cost. The ANZAC experience was undeniably a military defeat. Through 'heritagization', however, these experiences are redefined as turning points, climactic junctures in the destiny of peoples still fighting wars a long way from home on the basis of seemingly anachronistic and increasingly irrelevant imperial ties. These rebirths of identity are the achievement of heritage and, as such, become clearer and more obvious 'truths' about the war as time goes by. The potential of heritage remains latent until the necessary conditions or desire to operationalize it emerges (Lowenthal, 1996).

Conclusions

Although space constraints preclude the consideration of regimental museums, it is abundantly clear that changes are occurring in the perception of the Irish military experience. Although the official view of the Republic put aside the military history of the country, people have always remained conscious and interested in what their family members did in the wars. The symbolism of the Messines memorial signifies a move towards reconciliation with the past and was consciously integrated into the public discourse of the Northern Ireland Peace Process. 'Not since the early months of the Great War had everyone on the island of Ireland...been so unequivocally behind the same political objective' (Jeffrey, 2000, p. 3). The significance of the Messines battlefield may prove to be a latent basis for a new geographical 'reality': a 'unifying narrative of place'. To Irish identity, Messines may, in time, become redefined as a new starting point. It is only through the political dimensions of heritage that such a transformation could be achieved.

Acknowledgements

The author should like to acknowledge with gratitude the invaluable assistance provided through interviews with the following people: Mr T. Ball, Curator London Irish Rifles Museum, London; Mr T. Nelson, Museum Attendant, Royal Ulster Rifles / Royal Irish Rifles Museum, Belfast; Mr M. Hegan, Curator Royal Irish Regiment Museum, Ballymena; Major-General P. Nowlan, Chairman of the Military Heritage of Ireland Trust Ltd, Dublin.

References

Army Museums Ogilby Trust (1997), *Newsletter*, **Winter**, Salisbury.
Beckett, Ian. F.W. (1999), *Discovering British Regimental Traditions*, Shire Publications,

Risborough.

Bredin, A.E.C. (1987), *A History of the Irish Soldier*, Century Books, Belfast.

Connolly, S.J. (1997), 'Culture, Identity and Tradition: Changing Definitions of Irishness', in B. Graham (ed.), *In Search of Ireland: A Cultural Geography*, Routledge, London, pp. 43-63.

Douglas, Neville (1997), 'Political Structures, Social Interaction and Identity Change in Northern Ireland', in B. Graham (ed.), *In Search of Ireland: A Cultural Geography*, Routledge, London, pp. 151-73.

Duffy, Patrick. J. (1997), 'Writing in Ireland: Literature and Art in the Representation of Irish Place', in B. Graham (ed.), *In Search of Ireland: A Cultural Geography*, Routledge, London, pp. 64-83.

Farwell, Byron (1981), *Mr Kipling's Army*, W.W. Norton and Co, London.

Farwell, Byron (1989), *Armies of the Raj*, W.W. Norton and Co, London.

Gilbert, Martin (1994), *The Routledge Atlas of the First World War* (2nd ed), Routledge, London.

Graham, Brian (ed.) (1997a), *In Search of Ireland: A Cultural Geography*, Routledge, London.

Graham, Brian (1997b), 'Ireland and Irishness: Place, Culture and Identity', in B. Graham (ed.), *In Search of Ireland: A Cultural Geography*, Routledge, London, pp. 1-15.

Graham, Brian (1997c), 'The Imagining of Place: Representation and Identity in Contemporary Ireland', in B. Graham (ed.), *In Search of Ireland: A Cultural Geography*, Routledge, London, pp. 192-212.

Graham, Brian, Ashworth, G.J., and Tunbridge, J.E. (2000), *A Geography of Heritage: Power, Culture and Economy*, Arnold, London.

Hallows, I.S. (1991), *Regiments and Corps of the British Army, Arms and Armour*, Cassell, London.

Harris, R.G. (1999), *The Irish Regiments 1683-1999*, Sarpedon, New York.

Hobsbawm, Eric and Ranger, Terrance (1983) *The Invention of Tradition*, Cambridge University Press, Cambridge.

Hofstede, Geert (2001), *Culture's Consequences*, Sage, London.

Jeffrey, Keith (2000), *Ireland and the Great War*, Cambridge University Press, Cambridge.

Lowenthal, David (1996), *The Heritage Crusade and the Spoils of History*, Penguin, London.

McCrone, D., Morris, A. and Kiely, R. (1995), *Scotland - the Brand: The Making of Scottish Heritage*, Edinburgh University Press, Edinburgh.

Middlebrook, Martin and Mary (1994), *The Somme Battlefields*, Penguin, London.

Miller, Kenneth J.S. (1998), *British Military Museums: Visitor Development Opportunities*, Unpublished PhD Thesis, Napier University Business School, Edinburgh.

Miller, Kenneth J.S. (2001) 'Ethnographic Material in (British) Military Museums: A Conflict of Interests?', paper presented at the 60th Annual Meeting of the Society for Applied Anthropologists, Merida, Mexico, 31st March 2001.

Mitchell, D. (2000), *Cultural Geography: A Critical Introduction*, Blackwell, Oxford.

Murray, D. (2001), *Music of the Scottish Regiments*, Mercat Press, Edinburgh.

Museums and Galleries Commission (1990), *The Museums of the Armed Services. Report by a Working Party*, HMSO, London.

Schouten, F.F.J. (1995), 'Heritage as Historical Reality', in D. Herbert (ed.), *Heritage, Tourism and Society*, Mansell, London, pp. 21-31.

Sheen, John (1998), *Tyneside Irish*, Pen and Sword Books Ltd, Barnsley.

Smyth, William, J. (1997), 'A Plurality of Irelands: Regions, Societies and Mentalities', in B. Graham (ed.), *In Search of Ireland: A Cultural Geography*, Routledge, London, pp. 19-42.

Trevor-Roper, Hugh (1983), 'The Invention of Tradition: The Highland Tradition of Scotland', in E. Hobsbawm and T. Ranger (eds), *The Invention of Tradition*, Cambridge University Press, Cambridge, pp. 15-41.

Winter, Jay (1995), *Sites of Memory, Sites of Mourning: The Great War in European Cultural History*, Cambridge University Press, Cambridge.

Wood, S. (1993), 'At Their Country's Call. The Heritage of Soldiering in Scotland', in J.M. Fladmark (ed.), *Heritage, Conservation, Interpretation, Enterprise*, Donhead, London, pp. 271-81.

Chapter 8

A Place Called Nunavut: Building on Inuit Past

K.I.M. van Dam

Introduction

Nunavut (Figure 8.1) was created in 1999 following years of negotiation between the people living in the most northern part of Canada, the Inuit, and the Canadian government. These negotiations led to the settlement of land claims and the establishment of a new territory with its own government. In practical terms this means that the Inuit, 82 per cent of the population, have gained a form of self-government. Today the 26,750 inhabitants of Nunavut (based on the Census of 2001) live in approximately 26 small and medium-sized settlements across the territory (Nunavut Bureau of Statistics, 2002). In the formal economy in terms of employment and contribution to GDP, the service sector is most important. However, both the formal and informal economy depend for a large part on natural resources, varying from wildlife to minerals, gas and oil. The Inuit of Nunavut, like other indigenous peoples worldwide, have adopted sustainable development as a development strategy that would allow them to regain a balance between the environment and human activities. The creation of Nunavut has been discussed widely among social scientists, focussing on the importance for the identity of the Inuit (Merritt et al., 1987; Cameron and White, 1995; Soublière, 1999). Scientists and inhabitants alike agree that with this new territory, 'the Inuit of Canada's central and eastern Arctic are back in the driver's seat' (Soublière, 1999, p. 2).

It has been argued that it is important for communities to give meaning to a place (Massey 1995; Hall 1995; Rose, 1995; Allen et al., 1998; Crang, 1998). Sense of place is manifested in the formation of formal regions or nation states (Anderson, 1983) but is also expressed in more symbolic ways using cultural markers, tradition and heritage. In this respect, the establishment of Nunavut offers some interesting opportunities to discuss place identity. Despite the significant transformation in lifestyle, from being nomadic self-sustained hunters and gatherers to settled inhabitants of Canada, the historical-cultural traits of the traditional Inuit culture seem to be crucial in today's society. It is expected that in ascribing identities to Nunavut, these cultural traits of the hunter-gatherer society play a decisive role. At the core of this discussion is the role of place identity in a culture that   is generally believed to have different concepts of place than, for instance, the dominant concepts in the industrialized world. Analysing place identity within the context of an indigenous hunter-gatherer society is of particular interest in the light of discussions on the precise relation between cultural identity and place identity (Massey, 1995). The initial question that needs to be asked is what constitutes sense of place in a nomadic hunter-gatherer society? To be able to answer this question, the first part of this chapter discusses the general concept of identity, followed by a review of the notions of ethnic and cultural identity in the Inuit context and a discussion on place identity in hunter-gatherer societies. The other major issue discussed here concerns regional identity and, in particular, the nature of the regional identity that emerged during the formation of Nunavut.

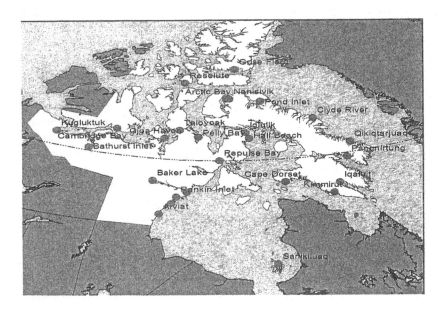

Figure 8.1 Nunavut

Dimensions of Identity

Since the 1970s, identity-based research in the Arctic has made notable advances (Dorais and Searles, 2001). The presence of several groups of indigenous peoples that live within the larger social and cultural context of a state contributed to a large number of studies in anthropology, sociology, linguistics and geography. These studies deal with identity from different perspectives, demonstrating that the concept of identity is multi-dimensional and complex. In general, identity is no longer defined as a fixed set of characteristics specific to a group of people, but is defined by social scientists such as Barth (1969), Anderson (1983) and Hobsbawm (1983) primarily as a social construction, as a means of relating to other people and other groups. In order to do so, characteristics that are perceived to be typical for the group are selected. These characteristics may be derived from culture, society, ethnic origin, geography and so on, but they are by no means fixed or their meanings uncontested. Various members of the group may ascribe different meanings to them, making identity a dynamic process. Identity also needs to be seen within the context of politics in which certain groups or actors make claims to achieve their goals. Another important general aspect of identity is that it always requires another person or group to be become manifest in making identity a relational concept (Crang, 1998). People either belong to a group and are similar to someone ('identifying with') or they are distinct and do not belong to the other group ('identifying against') (Rose, 1995). Identity is a concept that is composed of many different layers that are interconnected. Often a distinction is made between individual identity and collective identity, but others such as Légaré (2001) identify three levels of identity: individual identity (personal); social identity (gender; age); and group or collective identity. Collective identity then has two forms, namely cultural and territorial. Cultural identity refers to a particular cultural group (Inuit) and territorial identity to a political unit (Nunavut). Others base their analysis on a difference in scale moving from the individual via the collective and regional to the national identity level. To them, collective identity can be psycho-sociological (social class) in origin, or cultural, when expressing attachment to specific symbolic and material worlds (Dorais and Searles, 2001, p. 8). In this framework, ethnic identity can be seen as a form of collective identity but, once related to a group's hegemony over a territory, it could also be labelled as national (Dorais and Searles, 2001, p. 19). The analytic framework for this chapter is group or collective identity. In the same way that 'each person is ascribed at least several identities based on various group affiliations' (Dorais and Searles, 2001, p. 18), so group identities can be based on ethnic, cultural, geographical, social, political and other grounds. The unique blend of all these aspects is responsible for differentiating one group from another. Place identity or a sense of place is reviewed in a wider context than the geopolitical perspective of nations and/or territories.

Ethnic Identity

Ethnic identity can be defined as the consciousness which a group whose members are deemed to have the same geographic origins, phenotype, language or way of life has of its economic, political and culture distinctiveness in relation to other groups (Dorais and Searles, 2001, p. 19). It seems that origin in particular is a key element in ethnic identity. The Inuit of Nunavut are generally perceived as an ethnic group, sharing an ethnic identity that differentiates them from other groups or the broader Canadian society. The total geographical spread of Inuit is much wider: Inuit also live in Greenland, Alaska and Siberia. The geopolitical reality is that this ethno-cultural group is now part of several states in which they have a status, varying from being a minority group (Alaska) to forming the majority of the population in Greenland or Nunavut. Research focusing mainly on ethnic identity suggests that it is very often a politically charged concept. In the first studies on Inuit ethnic identity the relations between Inuit and the other group (Quallunaat or 'whites') were examined, using labels such as 'internal colonialism' and 'tutelage' (Brody, 1975; Paine, 1977). Later ethnic identity in particular has proven to be a strong force for political demands for self-determination. Aboriginal peoples use ethnic identity as a tool for furthering territorial and political claims.

Cultural Identity

Hall (1995, p. 176) defines culture as 'shared systems of meanings which people who belong to the same community, group or nation use to help them interpret and make sense of the world'. These meanings are embodied in the material and social world and are important for a sense of community. 'Culture is thus one of the principal means by which identities are constructed, sustained and transformed.' Usually the meanings and practices of culture have achieved a settled continuity over time and place and act as frame of reference of a tradition, which connects one's present mode of existence to the way of life of ancestors (Hall, 1995). Dorais and Searles (2001, p. 22) define cultural identity as a way of representing one's relation to nature, society and the supernatural. Manifestations of culture such as language, norms and values and worldview are used to constitute cultural identity. Studies on traditional Inuit culture in general and cultural identity in particular, give insight into which cultural traits are important for the cultural identity of Inuit. Inuktitut, the language spoken by the Inuit of Nunavut, is distinctive and as not all of the inhabitants of Nunavut speak English or French, Inuktitut becomes important for individual members of Nunavut and Nunavut society in general.

Much of the cultural identity of the contemporary Inuit is based on the traditional lifestyle of Inuit as hunter-gatherers. Having no arable land that could be used for farming, Inuit hunted and trapped wide areas. This lifestyle 'produced' clothing, tools, shelter (tents) and more. The direct and profound relation with the natural environment also led to a distinctive mythology and cosmology. Having no written language, spoken narratives were a crucial part in cultural and social life, in particular during the winter. Social and family organization was also adapted to life on the land, leading to specific forms of maintaining social order and handling conflict. The norms and values of Inuit society were directly linked to the subsistence lifestyle. Some do not regard subsistence lifestyle of hunting and gathering as being part of cultural identity but rather see it as a separate issue of identity (cf. Dorais and Searles, 2001). Because it is so integrated into all cultural traits this distinction seems only academic.

Place, Cultural Identity and Nomadic People

Most discussion on place and place identity in geography focus on the developed world although the concept of place may have different meanings in other cultures. Nomadic hunters-gatherer peoples such as the Aboriginal people in Australia, and pre-Columbian cultures in North America, base their concept of place on their attachment to the land and its natural features. People belong to the land and are defined by it (Massey, 1995).

> For most hunter-gatherer societies, the land has been an inextricable part of their lives; it provides not only their sustenance in the form of game, fish and vegetable foods, but also is at the foundation of their spiritual beliefs and social control mechanisms (Young, 1999, p. 320).

In general, nomadic peoples did not know property rights or bounded territories: territory boundaries fluctuated with the season and the size of the population. Although Inuit are no longer nomadic hunter-gatherers, their traditional subsistence lifestyle is still fundamental to much of their culture in which place was not necessarily equivalent to being settled. In most studies on Inuit identity their strong relationship to their environment becomes evident (c.f. Brody 1975; Nuttall 1998, 2001). In particular, studies that focus on the subsistence lifestyle indicate that Inuit are 'eco-centric': Inuit identity is dominated by the continual interaction with a living, transforming and sentient natural environment (Dorais and Searles, 2001, p. 21). When still nomadic, Inuit were living from the land: they are dependent on the land and its resources for their survival. The historical and contemporary use of areas, place names and a detailed knowledge of the natural world illustrate the strong bonds between people and place, environment, landscape and resources (Nuttall, 2001). In particular on the local level ideas of identity and attachment to a local area, or to '*nuna*', the land, figure prominently in everyday life. Although it is not the only way through which identity is constructed, locality and feelings of attachment to the local landscape are important markers of identity (Nuttall, 2001, p. 60). Thus Inuit do have a strong sense of place, directed towards the natural environment and the land itself. Nomadic societies occupy definite, if only diffusely, bounded social spaces over which they lay claim. Indeed, it was the strong relation to land that initiated the legal land claims and, in the Nunavut case, the creation of a territory.

Regional Identity in Nunavut

The North of Canada: Two Views

Although some Canadians claim the North is a symbol for Canada itself (Bone,

1992; Kassam, 2001; Grace, 2002), it is generally agreed that the North of Canada is formed by Nunavut, the Northwest Territories and Yukon Territory (Bone, 2000). Two perceptions dominate Canadian thinking about the North: the external perception of the Northern Frontier; and the internal perception of the Northern Homeland. To see the North as a frontier or resource region is a view originating from the south of Canada and it has long been the dominant view in Canadian public policy. As a frontier region, the North is an area in which to exploit natural resources, like oil, gas and minerals. In this view, the North is a harsh environment, but it contains enormous wealth. The area is often referred to as 'being empty'. Development is invasive and characterized by boom and bust cycles often caused by the global market. As a frontier region the North is mainly regarded as a supplier to southern markets. This relatively homogenous view of the North is 'exogenous' to the area and influenced by industrial capitalism and hinterland-metropolis relationships (Bone, 1992, 2000; Kassam, 2001). Today, it is often regarded as being 'myopic and simplistic' (Kassam, 2001, p. 437).

The North as a Homeland is a view of the North from the people who live and work there. This view is 'informed by thousands of years of indigenous use of the land and the sea' (Kassam, 2001, p. 433). The indigenous population is 2.8 per cent of the total population of Canada but in the northern territories the aboriginal share in the population is about 20 per cent in Yukon and 85 per cent in Nunavut. In general, the indigenous population is growing due to natural increase. The Homeland view is based on the importance of the traditional indigenous economy that relies on land and marine wildlife. Today, country food and economic activities related to hunting and trapping are still important for the indigenous economy in northern communities (cf. Usher, 1976; Wenzel, 1991; Duerden, 1992; Soublière, 1999). Besides economic relevance, most scholars agree that the harvesting of wildlife also contributes to a cultural and psychological well-being and sense of community spirit of indigenous people (Nuttall, 2000; Kassam, 2001). The Homeland view is associated with claims for land and self-determination by indigenous people. As Nuttall (2001, p. 379) argues, 'self-determination is the right to live a particular way of life, to practise a specific culture or religion, to use the language and the ability to determine the future course of economic development. Indigenous people distinguish themselves from other populations and they base their claims for a special legal status by reference to their special relationship to the environment and the importance of this relationship for the future of indigenous people'. The view of the North as a frontier has long been dominant. However, with the recognition of claims of indigenous peoples in the Arctic region over the past decades, it seems that the Homeland view has replaced the 'exogenous' frontier view (Osherenko and Young, 1989; Nuttall, 2000; Bone, 1992; Kassam, 2001). Nunavut Territory is generally seen as one of the most successful manifestations of this changing context (cf. Young, 1999; White, 2001). The discussion now turns to the creation of Nunavut and the associated development of regional identity.

The Territorial Shape: The Emergence of the Region

The establishment of Nunavut is based on the idea that the Nunavut Territory would better reflect the geographical extent of Inuit traditional land use and occupancy in the Canadian eastern Arctic, while its institutions would adhere to Inuit cultural values and perspectives (ITC, 1976, p. 15). Inuit wanted to control their political, social and economic agendas (c.f. Légaré, 2001, Nuttall, 2001). With the creation of Nunavut, the strong sense of place of the Inuit has been 'framed' in a regional or territorial context. This brings a new dimension to the sense of place discussion: one in which the identity ascribed to a region is central.

Paasi (1996) has identified four stages of the institutionalization of regions: the constitution of the territorial shape; symbolic shape; institutions; and the establishment of the territorial unit in the regional structure and social consciousness. Territorial shape refers to the localization of social practices through which regional transformation takes place and a territorial unit achieves its boundaries and becomes identified as a distinct unit on some scale of the spatial structure (Paasi, 1986). The formation of geopolitical borders is an important element of the institutionalization of a region. If boundaries do not exist, as was the case in Nunavut, they need to be defined. The formation of boundaries will often emerge as a result of a variety of cultural and political practices. If the geographical area is occupied by a distinct cultural group of people such as the Inuit, the production of geopolitical boundaries becomes a form of constructing and re-interpreting cultural space. Thus, space or the region's territorial shape is delimited by people's history, customs and tradition. It is not necessary for the cultural space and the boundaries of a region to correspond exactly since the interests of others (federal government, other Aboriginal groups) will be also have to be considered in the political arena. The aim of claimants to the emerging regions is then to establish political boundaries corresponding as much as possible to cultural boundaries. Through this whole process of boundary formation, boundaries create identity and are at the same time created through identity (Paasi, 1986; Légaré, 2001).

In Nunavut the basis for the territory is formed by the 'Land Use and Occupancy Study' (Freeman, 1976). This study was undertaken to prove aboriginal possession of land and to demonstrate Inuit land use (Merritt et al., 1987). Inuit hunters recalled their hunting journeys (referred to as 'living memories') to trace their territory. In addition to that, place names and historical cultural signs, camp sites, burial grounds and cairns were used to locate the extent of Inuit land occupancy. In the end, the maps showed the cultural space of the Inuit. This cultural space was used to determine the geographical extent of the claimed Nunavut region. According the federal government's Aboriginal Comprehensive Land Claims Policy, an Aboriginal group may hold land ownership and land management control over vast geographical areas, in exchange for proof of use and occupancy of the land. With the Land Use Occupancy Project in hand, the Inuit negotiators put forward a claim to a large part of the Northwestern Territories.

Regional identity is engraved on a group of people by actors who will subjectively use symbols and geopolitical borders to highlight the differences between one group and other groups. In doing so, they produce and reproduce regional identity. Although this chapter focuses mainly on the identity ascribed to Nunavut by internal actors, it is useful to consider briefly the other actors involved in the process of creating Nunavut. Claimants, such as the Inuit Tapirisat of Canada and, later, Nunavut Government, are the internal actors, representing the Inuit people. They negotiated with the policymakers and organized and informed the people of future Nunavut. Policymakers such as the federal government (Department of Indian Affairs and Northern Development) received the claim and had to adopt and implement solutions. The Aboriginal Comprehensive Land Claim Policy was used as a guideline for all actors. Other actors or stakeholders also played an important role in the process of shaping Nunavut. The cultural space of Nunavut is not uncontested: some land was used by other Aboriginal groups, such as the Dene-Métis and the Inuit of Northern Quebec, who have traditionally harvested on some portions of land within the claim area. The western boundary of Nunavut stretches over the cultural space of Dene Métis, whose traditional hunting grounds are now within Nunavut. Vice versa, the territorial shape of Nunavut does largely reflect the spatial extent of Inuit cultural space, but not entirely. The claimants had to compromise, for instance on the southern border where conflict would have arisen because of provincial integrity.

The Establishment of the Region

The claimants of the Inuit, the Inuit Tapirisat of Canada (ITC), put forward a Land Claim Proposal, which after years of negotiations let to the signing of the Nunavut Land Claims Agreement (NLCA) in 1993. This NLCA was a settlement based mainly on the ownership of land, but it included the 'commitment to create Nunavut Territory and Government on April 1, 1999'. This commitment was implemented the same year with the Nunavut Act: creating territory and government (Cameron and White, 1995; White, 2001). The last stage of Paasi's (1996) institutionalization of regions is the establishment of a region referring to the continuation of the institutionalization process after the region has an established status. It also refers to a specific regional identity in the spatial structure and social consciousness. Territorial units have specific 'structures of expectations' (Paasi, 1996, p. 35): expressing where the territorial unit has come from and where it is going. The motives of the Inuit to call for the territory and the organization of the territory after the creation are used here to provide insight in the ambitions of Nunavut. This is considered to be an important indication of the identity ascribed to the region. The main motives of the Inuit in creating Nunavut were: land ownership (the title to land, land claim); resource use and management (hunting rights and resource management); self-government and self-determination; and compensation

(Merritt et al., 1987; Cameron and White, 1995; Nuttall, 2000; White, 2001). The Inuit now own collective title to 350,000 km² and subsurface rights to 35,000 km². In the Nunavut Land-claim Agreement, it was also agreed that the Inuit would have more influence on matters related to land management and development of land. For this, so-called co-management bodies or Institutions of Public Government (IPGs) have been installed – federal government bodies that have to include Inuit (Merritt et al., 1987; DIAND, 2002). The IPGs manage wildlife, conduct environmental assessment and land use planning, and regulate the use of water. Finally, the creation of Nunavut Government (NG) established a new Territorial Government, over an area of almost two million km², which overlaps for the most part with the area of the land claim. The Nunavut Government is a public authority like all the other territorial governments in Canada, meaning that they are for all the inhabitants of Nunavut (26,750 people), and not for a specific group of the population (for example, the 23,000 Inuit). Compared to other territories in Canada, the NG has some special arrangements such as a decentralized form of government to spread the administration and employment over the territory. It has three official languages, instead of two: English, French and Inuktitut. There are no political parties, and government is based on consensus. The NG is also obliged to employ Inuit: targets and education programmes are part of the Nunavut Act.

The Symbolic Shape of Nunavut

According to Paasi (1996, p. 34), symbols are the keywords in the dominating story of a territorially based community. The formation of the symbolic shape of a region is expressed by specific symbols that are part of the symbolic ordering of space. During the construction of a region, symbols are established to emphasize the uniqueness and distinctiveness of the region and once boundaries are in place, symbols are used to reinforce this common collective identity. The naming of a place is one of the most fundamental activities in constructing place identities. The most important symbol is the name of the territorial unit or region. Names are part of the process of attaching meaning to one's surroundings and act as sources of information. Usually the name gathers together historical development, important events, episodes and memories and joins the personal histories of inhabitant to the collective heritage. In the case of Nunavut, it is the first time the region gets a name that appears on maps. Before that, the area was part of another region, the Northwest Territories, which made sense to Canadians, but did not relate to the Inuit. For the Inuit, the naming has been very important, although they had long ago given the land a name: simply, Nunavut, 'our land'. In traditional Inuit society 'nunavut' was used to mean a land familiar to Inuit hunting parties. The transfer of power to the Inuit population has also led to changes in the names of the settlements. Nunavut's 26 settlements have a relatively short history. Most of them grew from trading posts or missionary posts at the beginning of the twentieth century. All the settlements in the area had names that were English-Canadian in origin. Since most of the settlements were established near places or areas that were of importance for Inuit, they usually had an Inuit name for the place as well. Since the 1980s, the inhabitants of Nunavut have started to change the names of the settlements into Inuit-names and after the creation of Nunavut, these new names appear officially on the maps. The new capital of Nunavut, Iqaluit, was known as Frobisher Bay, after the explorer Martin Frobisher. Now the place is called Iqaluit meaning 'place of fish'. A lot of places in Inuktitut refer to animals (fish', 'caribou', 'birds'), geographical features ('cove', 'mountains'), natural events ('where there is no daylight', 'place that never melts') or the presence of people ('where many people arrive') (*Canadian Geographic*, 1999; *Nunavut Travel Planner*, 2002). Apart from these official place names, the Inuit have names for most parts of the land, as is illustrated on land use maps of the Land Use Occupancy Project. More recently, a map drawn by Inuit elders in a study on caribou also proved that the contemporary landscape of Nunavut is still dotted with names (Thorpe et al., 2002).

As Paasi (1996) argues, the increased use of territorial symbols and signs is linked to the emergence of institutions that socialize people into territorial memberships. These memberships connect the inhabitants with the symbols of the region in various practices and at the same time demarcate the 'Other'. These, often abstract, symbols are instrumental in that they serve to evoke powerful emotions of identification with territorial groups and can generate action. Symbols are borrowed from elements of a group's cultural identity and are often the idealized version of cultural traits or cultural markers. Légaré (2001) examines the most reoccurring symbols of identity in the region that rest upon three forms of manifestation: rituals, pictorial graphics and names. It is assumed here that identification with a region is translated in using the name in other contexts. Besides the naming of 'Nunavut' for the territory and 'Nunavummiut' for the inhabitants, the use of the name of the region by institutions has increased. In 1970 the name was unknown for institutions; in 1995, more than 52 companies and institutions included this spatial terminology as part of the institutions' name identification (Légaré, 2001). Advertisements and the *Business Directory of Nunavut* show a number of firms have started to use 'Nunavut' in their name. Besides that, other institutions such as the co-management bodies and education facilities (for example, Nunavut Research Institute) also use 'Nunavut' in their name, as does the newspaper *Nunatsiaq News*. In 1978 'Nunavut' was found 57 times in all printed articles and the word 'Nunavummiut' was absent. In 1998, Nunavut appeared 556 in a year and 'Nunavummiut' occurred 106 times (Légaré, 2001). What symbols are chosen to represent the region, also carries a message on the identities ascribed to it. As far as the pictorial symbols or logos are concerned, all sorts of elements from Inuit culture and environment have been used. One of the most popular symbols is the inukshuk. This is a cairn used by Inuit hunters as a reference point or navigator, now used by Inuit organizations and in tourist brochures in logos and emblems. The inukshuk is a clear example of how a cultural-historical element has turned into a symbol to stress the distinctiveness of the region and its people. Other symbols are used in much the same way: the drum (in the logo of the Tourist Board), the polar bear (Nunavut Government, licence plate), and people (Inuit or inuk) (ITC). Finally, the official symbolism in the flag and the coat of arms again use the inukshuk, caribou, narwhal and oil lamp. As far as rituals are concerned, Nunavut has established the Nunavut holiday, on the day (July 9) the Nunavut Political Agreement received royal assent. Indeed it seems that the formation of symbols in Nunavut is largely based on regional, cultural and physical elements of distinctiveness: the Arctic climate and ecology, and the Inuit traditional economy of harvesting.

Conclusion

It summary, therefore, for the people living in the area, Nunavut identity is based on the people and the community, along with Inuit culture and the land. To them, Nunavut is not an empty and cold land, but a land with rich resources on land and sea. The dominant view of southern Canadians has been contested successfully, resulting in the establishment of Nunavut Territory. This shift in the power balance has changed the context of Nunavut, and the identities that are ascribed to the place. Still, the identity ascribed to the area that is now Nunavut continues to be dynamic and contextualized. For instance, it is important to realize that so far, the identities ascribed to Nunavut have been analysed within the context of government, Inuit stakeholder organizations and businesses. Some suggest this group is an élite (Mitchell, 1996), not necessarily reflecting possible alternative identities of the inhabitants of Nunavut. Another interesting issue for further research is the nature of the future identities that will be ascribed to Nunavut, both internally and externally. The now dominant homeland-identity might be challenged by different actors within and outside the region or by economic and demographic development. Apart from Nunavut as a homeland, there might also be a Nunavut as the pristine wilderness and even Nunavut as a resource region may still exists.

References

Allen, J., Massey, D. and Cochrane, A (1998), *Rethinking the Region*, Routledge, London.

Anderson, B. (1983), *Imagined Communities* Verso, London.

Beluga Adventures (2003), *Reizen in (ant) Arctische Gebieden* Beluga Adventures, Pernis.

Bhattacharyya, D.P. (1997), 'Mediating India: An Analysis of a Guidebook', *Annals of Tourism Research*, **24**, pp. 371-89.

Bone, R.M. (1992), *The Geography of the Canadian North*, Oxford University Press, Don Mills.

Bone, R.M. (2000), *The Regional Geography of Canada*, Oxford University Press, Don Mills.

Brody, H. (1975), *The People's Land: Whites and the Eastern Arctic*, Penguin, Harmondsworth.

Cameron, K. and White, G (1995), *Northern Governments in Transition: Political and*

Constitutional Development in the Yukon, Nunavut and the Northwest Territories, Institute for Research on Public Policy, Montreal.

Canadian Geographic (1999), *Nunavut*, Royal Canadian Geographic Society, Ottawa.

Conference Board Canada (2001), *Nunavut Economic Outlook*, Conference Board Canada, Ottawa.

Crang, M. (1998), *Cultural Geography* Routledge, London.

Dorais, L.J. and Searles E. (2001), 'Inuit Identities', *Études Inuit Studies*, **25**(1-2), pp. 17-35.

Duerden, F. (1992), 'A Critical Look at Sustainable Development in the Canadian North', *Arctic*, **45**(3), pp. 219-25.

Freeman, M.R. (ed.) (1976), *Inuit Land Use and Occupancy Project*, INA publication.

Grace, S.E. (2001), *Canada and the Idea of North*, McGill-Queen's University Press, Montreal and Kingston.

Haartsen, T., Groote, P and Huigen P.P.P. (eds) (2000), *Claiming Rural Identities: Dynamics, Contexts and Politics* Van Gorcum, Assen.

Hall, S. (1995), 'New Cultures for Old', in D. Massey and P. Jess (eds), *A Place in the World?*, Oxford University Press, Oxford, pp. 175-215.

Hopkins, J. (1998), 'Commodifying the Countryside: Marketing Myths of Rurality', in R. Butler, C.M. Hall and J. Jenkins (eds), *Tourism and Recreation in Rural Areas*, Wiley, New York, pp. 139-56.

Jochim, M.A. (1976), *Hunter-Gatherer Subsistence and Settlement*, Academic Press, New York.

Kassam, K.S. (2001), 'North of 60∞: Homeland or Frontier?', in D. Taras and B. Rasporich (eds), *A Passion for Identity: Canadian Studies for the 21ˢᵗ Century*, Nelson Thomson, Scarborough, 4ᵗʰ ed, pp. 433-55.

Kruk, E. (1998), 'Nunavut Hoopt op Pooltoerist', *Geografie*, **7**, pp. 8-10.

Lanting, E. (1995), *Sumut – Waarheen – Groenlandisering en het Dagelijkse Leven in het Upernavk District, Groenland*, PhD Thesis, University of Groningen.

Légaré, A. (2001), 'The Spatial and Symbolic Construction of Nunavut: Towards the Emergence of a Regional Collective Identity, *Études/Inuit/Studies*, **25** (1-2), pp.141-68.

Massey, D. (1995a), 'Introduction', in D. Massey and P. Jess (eds), *A Place in the World?*, Oxford University Press, Oxford, pp. 1-4.

Massey, D. (1995b), 'The Conceptualization of Place', in D. Massey and P. Jess (eds), *A Place in the World?*, Oxford University Press, Oxford, pp. 45-86.

Massey, D. and Jess, P. (eds) (1995), *A Place in the World?*, Oxford University Press, Oxford.

Merritt. J., Fenge, T., Ames, R. and Jull, P. (1987), *Nunavut: Political Choices and Manifest Destiny*, Canadian Arctic Resources Committee, Ottawa.

Nunavut Bureau of Statistics (2002), *Population Counts from the 2001 Census*, Government of Nunavut, Iqaluit.

Nunavut Tourism (2002), *Canada's Arctic Nunavut Travel Planner*, Government of Nunavut, Iqaluit.

Nuttall, M. (1998), *Protecting the Arctic: Indigenous Peoples and Cultural Survival*, Harwood Academic Publishers, Amsterdam.

Nuttall, M. (2000), 'Indigenous Peoples, Self-determination and the Arctic Environment', in M. Nuttall and T. Callaghan (eds), *The Arctic: Environment, People, Policy*, Harwood Academic Publishers, Amsterdam.

Nuttall, M. (2001), 'Locality, Identity and Memory in South Greenland', *Études/Inuit/Studies*, **25**(1-2), pp. 319-38.

Osherenko, G. and Young, O. (1989), *The Age of the Arctic: Hot Conflicts and Cold Realities*, Cambridge University Press, Cambridge.

Paasi, A. (1996), *Territories, Boundaries and Consciousness: The Changing Geographies of the Finnish-Russian Border*, Wiley, Chichester.

Paine R. (ed.) (1977), *The White Arctic: Anthropological Essays on Tutelage and Ethnicity*, University of Toronto Press, Toronto.

Pater, B. de, Groote, P. and Terlouw K. (2002), *Denken over Regio's,* Coutinho, Bussum.

Rose, G. (1995), 'Place and Identity: A Sense of Place', in D. Massey and P. Jess (eds), *A Place in the World?*, Oxford University Press, Oxford, pp. 87-132.

Shurmer-Smith, P. (ed.) (2001), *Doing Cultural Geography*, Sage, London.

Soublière, M. (ed.) (1998), *Nunavut Handbook*, Nortext Multimedia, Iqaluit.

Soublière, M. (ed.) (1999), *Nunavut 99*, Nortext Multimedia, Iqaluit.

Thorpe, N., Hakongak, N. and Eyegetok, S. (2002), *Thunder on the Tundra*, Tuktu and Nogak Project, Victoria.

Usher, P. (1976), 'Evaluating Country Food in the Northern Native Economy', *Arctic*, **29**(2), pp. 103-20.

White, G. (2001), 'Government Under the Northern Lights: Treaties, Land Claims and Political Change in Nunavut and the Northwest Territories', in D. Taras and B. Rasporich (eds), *A Passion for Identity: Canadian Studies for the 21ˢᵗ Century*, Scarborough, 4th edition, pp. 457-76.

Young, E. (1999), 'Hunter-Gatherer Concepts of Land and its Ownership in Remote Australia and North America', in K. Anderson and G. Fay (eds), *Cultural Geographies*, Longman, Addison, pp. 319-39.

Chapter 9

Conflict Commemoration Amongst Protestants in Northern Ireland

Catherine Switzer

Introduction

> No monument was needed to keep alive the memory of those splendid
> men who so unflinchingly carried out their duty to King and country.
> Their memories were enshrined in their own hearts, and would never
> fade; but it was a fitting thing that an outward and permanent symbol of
> their gratitude should be erected in their midst, so that those passing by
> might be reminded of the part played by the men of the district in the
> war... the memorial would always be their most cherished possession.
> J. Taylor Stronge, Chairman of Donaghadee Urban District Council, at
> the unveiling of Donaghadee War Memorial (*Belfast Newsletter*, 2 July
> 1926, p. 7).

The war memorial in Donaghadee, a small seaside town on the northern edge of
County Down in Northern Ireland, was only one of a large number of public
memorials erected in the years following the Great War of 1914-18 to commemorate
those who had not returned from that conflict. The period following the Armistice
in 1918 marked the beginning of widespread public conflict commemoration in
Northern Ireland, as the erection of war memorials was combined with the invention
of new commemorative rituals. Yet despite the service of Ulstermen of all political
and religious persuasions in the British Army in that conflict, the dominant memory
of the war as it developed in the North of Ireland emphasized a particularly
Protestant and unionist version of events. As historian Keith Jeffery has observed:

> In the North the commemoration of the war became overwhelmingly an
> opportunity to confirm loyalty to the British link and affirm Ulster's
> *Protestant* heritage (Jeffery 2000, p. 131, emphasis in original).

Although the outward forms of commemoration have not changed greatly since the 1920s, the meaning of the rituals has been broadened and the memorials rededicated to include a wider range of conflicts. In Northern Ireland this refers most often to World War II (1939-45) and the more recent Troubles. More specifically, public commemoration of members of the security forces killed during the Troubles draws on elements of world war commemoration, and raises important issues about the purposes to which commemoration is put. Using examples drawn from across Northern Ireland, this chapter illustrates some of the ways in which conflict commemoration has evolved over time. In doing so, it has two primary aims. The first is to provide a broad outline of the ways in which the World Wars and the Troubles have been memorialized in Northern Ireland. The second is to examine two particular aspects of public commemoration of members of the security forces which draw on pre-existing war commemoration, and to explore some of the consequences of these linkages.

Studying Commemoration

The Great War is a suitable starting point since it provided the impetus for the first widespread commemoration of soldiers killed in warfare. In the intervening years, its memorials have been used to commemorate the dead of other wars. In many ways, domestic war memorials have their genesis in the confusion and mire of Great War battlefields. Difficulty in the identification and recovery of corpses led the War Office in 1915 to ban the repatriation of the bodies of dead British servicemen. Instead, official memorial landscapes of military cemeteries and monuments were created on battlefields under the auspices of the government-funded Imperial War Graves Commission (Longworth, 2003). Domestic memorials were therefore considered necessary in the absence of the corpses which had traditionally provided a focus for the traditions surrounding death and burial.

The impact of the Great War on the north of Ireland and the ways in which it was subsequently remembered will be dealt with later, but in a broader context, the renewed interest in the Great War has during recent years produced an extensive international literature on commemorative practices. The products of this scholarship are now so numerous as to constitute what Inglis (1998, p. 8) has termed a, 'vigorous branch of cultural history'. In particular, practices of commemoration and war memorials have become subjects of study in their own right. The literature can be divided into two principal categories.

Briefly, the first, as exemplified by Mosse's book *Fallen Soldiers* (1990, pp. 6-7), sees memorials and commemoration as being tied up with the workings of nationalism, which sought to, 'draw the sting from death in war and emphasize the meaningfulness of the fighting and sacrifice'. This was important, 'above all for the justification of the nation in whose name the war had been fought'. Mosse dubs this, 'the myth of the war experience', a myth which refashioned the memory of the war into a, 'sacred experience which provided the nation with a new depth of religious feeling, putting at its disposal ever-present saints and martyrs, places of worship and a heritage to emulate'. This myth is symbolized in three tangible ways: through military cemeteries, memorials to the war, and the commemorative ceremonies held for the dead.

Secondly, memorials, in their initial stages at least, can be interpreted as providing a focus for grief and a place, 'where people could mourn. And be seen to mourn ... [providing] first and foremost a framework for and legitimation of individual and family grief' (Winter, 1995, p. 93). Although Winter finds merit in Mosse's work, he contends that an approach centred on the politics of commemoration may neglect the reason behind the memorials' original purpose. He notes the difficulties posed by language, but contrasting French *monuments aux morts* with English war memorials and the German *kriegerdenkmal*, he finds that, despite the memorials' invitation to recall more than the loss of life and bereavement of war, these remain the central facts:

> ... [T]hese monuments had another meaning for the generation that passed through the trauma of the war. That meaning was as much existential as artistic or political, as much concerned with the facts of individual loss and bereavement as with art forms or with collective representations, national aspirations, and destinies (Winter, 1995, p. 79).

The categories above are not, of course, mutually exclusive since memorials embody a whole range of meanings about power, social status and ideology (Inglis and Phillips, 1991). As Moriarty (1997) has argued, they can be regarded as composite sites at which the commemorative element is only one of many possible readings.

Both approaches may also contribute to an understanding of memorials in landscape and in place. Thinking about these existing sites from a historian's perspective, Inglis (1998) sees war memorials as *Sacred Places*, albeit not in a conventional religious sense. Since they constitute a central part of what Young (1993, p. 2) terms, 'the objects of a people's national pilgrimage', it is possible to think about memorials as sacred in a secular sense through their importance to the continued existence of nationalism, and it would be tempting to link them to Mosse's (1990) cult of the fallen soldier. As Winter points out, to do so would be to ignore their function as a focus for grief and mourning. For Inglis (1998, p. 11), this is where the idea of the sacred originates:

> the statue on its pedestal did stand for each dead man whose body, identified or Missing, intact or dispersed, had not been returned ... That made its site holy ground.

Thus war memorials can be regarded as sites of complex and sometimes contradictory meanings, which can be located simultaneously in the context of nationalism, or in the context of grief and mourning.

Northern Ireland, War and Commemoration

Here I turn to examine the impact of three conflicts on Northern Ireland: the Great War, World War II and the more recent domestic Troubles, and the ways in which each conflict has been memorialized.

The Great War (1914-1918)

Although in many ways the patterns of memorialization and commemoration as they developed in Northern Ireland were unremarkable in comparison with other regions of the British Empire, the Irish experience of the Great War was very much rooted in the internal political situation of the immediate pre-war period. In the summer of 1914, unionists and nationalists in Ireland were on the verge of civil war over the implementation of Home Rule by the British Government. Two paramilitary organizations were formed to contest the issue. The Ulster Volunteer Force intended to resist the implementation of Home Rule, by force of arms if necessary, while the Irish Volunteers were equally determined to enforce it. In the event, the conflict was postponed when war was declared and men from both paramilitary groups joined the British army, Ulster Volunteers making up the majority of the 36th (Ulster) Division, and many Irish Volunteers joining the 16th (Irish) Division. A third Division, the 10th (Irish) was religiously mixed (Orr, 1987), while many thousands of Irishmen served in the Dominion armies. Despite this diversity of service and experience, in the North of Ireland the war became condensed into a single event: the charge of the Ulster Division on 1 July 1916, the opening day of the Battle of the Somme. Around 5,500 men of the Division became casualties over the first three days of the battle, of which around 2,000 were killed on the first day alone (Middlebrook, 1971).

In the wake of the Armistice in November 1918, committees acting on behalf of local communities began to plan how to mark the war in a permanent way, and in many cases their solution involved monumental sculpture. The majority of war memorials in Northern Ireland date from the 1920s and early 1930s, two of the earliest being unveiled in 1922 in Portrush and Coleraine. Although one of the latest in that first phase was dedicated in Newtownards in 1934, a number have been erected more recently. From 1919 onwards, Armistice Day – November 11th – was established as a day for remembering the war dead, while the form of commemorative ceremonies as they still take place today was largely established in the immediate post war years. Some aspects of Great War commemoration in Ireland have recently been explored by Johnson (2003) and Loughlin (2002).

World War II (1939-1945)

So by the advent of World War II, the pattern had already largely been set, the memorial sites chosen and the practices of commemoration established. Northern Ireland's experience of World War II differed greatly from that of the Great War

25 years before. Military losses were fewer, and spread between units and over time, so there was no equivalent to the Somme this time, and no unit of the armed forces to identify specifically with Ulster or Northern Ireland. At home, four nights of German air raids on Belfast in 1941 resulted in a death toll of around 1,100 people, while another 100,000 were made temporarily homeless (Barton, 1995). Although the experience of the Blitz has inspired quite extensive memorials in other British cities, perhaps most notably in Coventry, this tragic event in Belfast's history has been more or less completely effaced from the memorial landscape. Existing war memorials were often rededicated and their inscriptions altered to include the dates of the war and a second list of names, although not all were updated. In terms of memorials, then, World War II in Northern Ireland appears as a continuation or even as a lesser version of the Great War; its dates and the names of the military dead listed below those of the previous war where the focus remains. Civilian deaths are not visible at all.

The Troubles

Although my focus here will be on the memorials to members of the security forces which have been placed adjacent to existing public war memorials, it is worth briefly considering the broader context of Troubles commemoration. Kenneth Bloomfield observes in his report *We Will Remember Them*, that commemoration in Northern Ireland, 'can too easily take on a confrontational quality', providing an impetus for further violence (1998, p. 11). Bloomfield headed a Commission established by the British Government which was charged with investigating ways to, 'recognise the pain and suffering felt by victims of violence arising from the troubles of the last 30 years, including those who have died or been injured in the service of the community' (1998, p. 8). The Commission recommended that, following the provision of a wide range of practical support for victims and the bereaved:

> at the appropriate time, consideration should be given to a Northern Ireland Memorial in the form of a beautiful and useful building within a peaceful and harmonious garden ... such a project should be called simply 'the Northern Ireland Memorial' (Bloomfield, 1998, p. 51).

The Commission's conception of a utilitarian memorial is driven perhaps by the idea that something positive should come out of the conflict. The memorial would also seem to honour *all* victims together, regardless of their background. In contrast, the majority of existing Troubles memorials tend to lack utilitarian functions. They also tend to focus on individuals as members of particular groups. In recent years, civilian victims of several incidents, including the 1972 bombing in the village of Claudy, have become the subject of public memorials. Bloody Sunday (also 1972), an event still mired in controversy, is marked through a memorial and commemorative marches, although the 'ownership' of the event itself remains contested (Dunn, 2000). Paramilitary groups have been active in commemorating their own members through wall murals, memorials and marches within their areas of influence (Jarman, 1997).

My more specific concern here lies with some of the ways in which members of the security forces killed during the Troubles have been commemorated. Although in general usage, the term 'security forces' is ill-defined; it is used here to refer to members of the Ulster Defence Regiment (UDR) and its successor the Royal Irish Regiment (RIR), along with the Royal Ulster Constabulary (RUC), now known as the Police Service of Northern Ireland (PSNI).

Established in 1970 partly as a replacement for the controversial 'B' section of the Ulster Special Constabulary (USC), the UDR was an infantry regiment of the British army, which was recruited from and operated only in Northern Ireland. Members could be full-time soldiers, although the majority worked on a part-time basis. In 1992 the regiment was merged with the Royal Irish Rangers to become the Royal Irish Regiment (Ryder, 1992). The RUC, which was established in 1922 after partition as the police force of Northern Ireland, received the George Cross (GC) in April 2000, the highest honour that can be awarded to civilians in the UK. In a controversial move, the force was subsequently renamed and re-badged under the recommendations of an independent commission lead by British politician Chris Patton and has since been known as the Police Service of Northern Ireland (Ryder, 2000). The overwhelming majority of members of both the UDR and the RUC were Protestants, despite efforts on the part of both organizations to recruit Catholics.

The landmark publication, *Lost Lives* (McKittrick et al., 1999), which is in itself a form of memorial to those who have died in the Troubles, finds that, of a total of 3,636 deaths up to the summer of 1999, 1012 were members of what can be loosely termed the security forces. Of these, 503 were members of the 'mainland' armed forces, 206 belonged to the UDR or its successor the RIR, and 303 were members of the RUC or its full-time and part-time reserve (RUCR). Partly as a result of their spread between various branches of the armed forces and the police, no overall official memorial exists to commemorate these deaths, although as some of the entries in *Lost Lives* indicate, their names may appear on regimental or other unit memorials. Perhaps the most comprehensive memorial so far, albeit located outside Northern Ireland, is the Ulster Ash Grove at the National Memorial Arboretum in Staffordshire, England. Dedicated in September 2003, the memorial honours members of mainland units, the UDR, RIR, RUC and the Prison Service. It is focused around six large boulders, one brought from each county of Northern Ireland, along with a central stone and seats hewn from Mourne granite (National Memorial Arboretum, n.d.). The memorial aspect consists of more than 1,000 ash saplings, each of which bears a tag carrying the name, regiment or unit, and date of death of one individual.

Within the boundaries of Northern Ireland, as Leonard (1997) notes, memorials to members of the security forces, often in the form of plaques, are generally found on army and police premises. Such memorials are therefore seen only by other members of the security forces, or by family members who are able to gain access to them and are often present when the memorials are unveiled (See, for example, *Police Beat* 1980; 1981; 1982; 1984). The RUC GC Garden at PSNI Headquarters in Belfast, opened by Prince Charles in September 2003, although the most extensive RUC memorial, is no different since it can be viewed only by appointment. Individuals may also be commemorated in a wide variety of other ways, for example through the dedication of items of church furniture in their memory, or the presentation of items, such as a painting or clock, to the station at which the individual was based.

Continuity of Commemoration

My concern here, however, is with memorials which are in public places and, more specifically, those which are connected with public war memorials. I have shown above how sites originally created to facilitate commemoration of the Great War have been rededicated to include other conflicts, but also how the Great War remains the dominant feature of commemoration. The following discussion examines the linkages that exist between commemoration of security force personnel and that of the war dead. Two specific subjects will be discussed: the location of memorials to members of the security forces; and the inscriptions on the memorials.

Locating Memorials

Although the majority of memorials to members of the security forces are found inside army bases and police stations, a number of memorials do exist in public places. In a few cases, the names of members of the UDR have been added to the existing war memorial, but this is rare. More often, a small memorial or plaque has been placed alongside the existing war memorials, although several Troubles' memorials stand independently. In contrast to memorials inside security premises, the security forces themselves did not erect those in public places.

The majority of existing war memorials in Northern Ireland stand inside a particular area which is set aside for the express purpose of commemoration and is clearly defined by a wall or fence. I have suggested above that it might be possible to see war memorials as constituting in some sense 'sacred' space. As spaces set aside from everyday use, which are used only on particular dates of the year, it is certainly not difficult to see them as distinct from, and standing outside of, the patterns of day-to-day life. As sites where regular commemorative services have been held for 70 years or more, existing war memorials also benefit from the authority bestowed on them by tradition. Kenneth Foote has described how the sanctity of particular sites already established as important may be reinforced through time by the construction of additional memorials, a process he calls 'symbolic accretion' (1997, p. 231).

One particularly extensive example of this can be found in the Cregagh area located on the edge of Belfast in Castlereagh. The memorial site is at the centre of an area of housing known as the Colony, built during the 1920s specifically to house veterans of the Great War, and bearing streetnames taken from the Somme battlefield: Bapaume, Thiepval, Hamel, Picardy and Somme. These names, particularly the evocative Thiepval, are an implicit reference to the 36th (Ulster) Division; no streets in this area are named after Guillemont or Guinchy, places also on the Somme but associated with the 16th (Irish) Division. Given this, it is noteworthy that the Great War memorial takes the form of a rough Celtic cross, which might be seen as an 'Irish' symbol, and was rarely used in Great War memorials in Northern Ireland. The cross was first unveiled in 1929, only to be re-dedicated in 1932 during a visit to Northern Ireland by the then-Prince of Wales. The memorial as it existed at that time was therefore already a site of some complexity.

The site is now used as a centre for the public commemorative activity of Castlereagh Borough Council, which has added a number of further memorials during the 1990s. The Celtic cross remains the central feature, but is now flanked by stones dedicated to the USC and the UDR. These also record the grant of the freedom of the Borough to their respective Associations. The rear wall of the memorial area features plaques dedicated to (from left to right) the RUC, RUCR and USC; named residents of the area killed in World War II; 'all innocent victims of terrorism who lived in the Borough of Castlereagh'; and members of the UDR.

It is obvious that, on top of its original complexity, the Colony is now the site of multiple memorials to multiple conflicts. The Great War remains central, not least through the dominating positioning of the Celtic cross, but also by virtue of the surrounding street names with their Great War resonance. The additional memorials, however, create a situation in which other soldiers, and indeed police officers, may be seen as equivalent to the heroes of the past. There is an implied temporal continuum, suggesting an unbroken line of service from the Great War through a succession of conflicts to the present day. There is more than a small degree of contradiction in this, not least because the concept of the Troubles as a war is more often associated with the Irish republican movement. If, for unionists, the Troubles did not constitute a war, then a further question is raised: is a war memorial the right place to commemorate those who have died as a result of terrorist activity? It might also be observed that, visually, World War II appears equivalent to the Troubles, or to go further, that it is of less importance than the domestic conflict which has inspired five memorials.

There is a further aspect to this process of symbolic accretion. Foote (1997) sees it as serving to sanctify further sites which were already important, but it is also possible to argue that security force memorials cause this process to work in reverse, to the detriment of the original memorial. Moving away from

Cregagh, the memorial in Portadown serves as an example of this converse process. As has already been stated, commemoration of the war dead is widely seen as the preserve of the Protestant community, although for many unionists this has occurred by default through the non-participation of Catholics, rather than through any exclusionary element within commemoration itself.

Sited in the centre of a town which, as Jeffery (2000, p. 133) puts it, has a, 'grand old (and continuing) reputation for sectarian strife', the memorial in Portadown was unveiled in 1925 and is dedicated to 'the glorious memory of the men of Portadown and its neighbourhood'. The memorial makes no distinction between the men it names, although as they are listed by street, or by townland outside the urban area, their religious background might be inferred from where they lived, or indeed by their names. The commemorative ceremonies focused on the memorial might be seen as the property of the Protestant community, but the memorial itself honours all dead servicemen.

This memorial has been joined by a small obelisk which carries the names of six members of the local UDR battalion who died violently during the Troubles. Placing a memorial to members of a predominantly Protestant regiment of the British army alongside the existing war memorial could be seen as an active step towards claiming the memorial as Protestant, thus challenging the unionist assertion that war commemoration is only Protestant, or unionist, because Catholics do not attend. Such memorials could also threaten to disrupt current attempts to use the common Irish experience of the Great War as a focus for reconciliation in present-day Northern Ireland.

Inscriptions

The inscriptions on war memorials can provide a productive area for study since they define meaning. Memorials created *during* the Great War are part of what Winter (1995, p. 85) terms:

> the initial phase of commemorative art, in which the glorification of sacrifice was expressed in a deliberately archaic language, the cadences of knights and valour, of quests and spiritualized combat.

Fussell (1975, pp. 21-2) notes that although this, 'raised, essentially feudal' language eventually became a casualty of the war, 'its staying power was astonishing'. He explores its use in the literature generated by the war, but many examples of this 'high' diction, albeit slightly toned down, were engraved into memorials to instruct viewers on how to understand the war. The memorial in Gilford, Co. Down, is dedicated to the 'Memory of our fallen heroes,' while 'Glory to the Dead' is exhorted in Armagh. King (1998) places emphasis on the didactic character of Great War memorials and commemorative ceremonies, which taught that the dead should be respected and their deeds should be valued. Virtues were attributed to the dead in order to justify honouring them. Death on the field of battle was constructed as honourable, and those who fell performing gallant deeds were deserving of glory, despite the fact that for many of them their deaths were presumably inglorious. The poet Michael Longley (1997, p. 122, emphasis in original) has written that this language 'encourages us *not* to remember how shrapnel and bullets flay and shatter human flesh and bones'.

There are interesting parallels and divergences between the inscriptions on Great War memorials (which, as I have already stated, now also stand for World War II) and those which have been erected alongside them to commemorate members of the security forces. A number of security force memorials make use of sentiments associated with existing war commemoration, such as 'lest we forget' and 'we will remember them'. Security force memorials are, however, notably devoid of 'high' diction with its implications of honour and glory for the fallen. Those named are not generally asserted as having died for any of the 'big words' which often appear on Great War Memorials: God, King, Country, *Pro Patria Mori*, the last phrase perhaps better known today as Wilfrid Owen's 'old lie'. Instead inscriptions tend to employ less elevated language, a number moving in an entirely different direction. Whereas the inscriptions on Great War memorials act to hide the brutal truth of death in war, some inscriptions on security force memorials encourage the reader to consider the way in which those commemorated died. A number assert that those named were 'Murdered in the execution of their duty', a phrase with multiple meanings which hinge on the different ways in which execution can be carried out. Police officers and soldiers, it is implied, execute their duty, while terrorists, although not mentioned explicitly, execute people. Other memorials use the word 'killed' as an alternative to 'murdered', but this does not constitute a move much further towards 'high' diction. The apparent reluctance to use the word 'fallen' prompts a somewhat uncomfortable parallel with memorials in nineteenth-century Australia, of which Inglis (1998, p. 24, emphasis in original) observes that, 'Nobody was said to have *fallen* in battle against the natives'.

Such inscriptions challenge King's assertion that the didactic function of memorials and commemoration work through the attribution of virtues to the dead. The memorials do not generally denote members of the security forces as glorious or deserving of honour because of what they have done or the manner of their deaths. Instead, they are constructed as people who did their duty and while doing so were murdered, although the precise identity of the guilty party is left to the reader's interpretation. It may be that the proximity of an existing war memorial necessitates a toning-down of language; the wording of memorials which 'stand alone' is less circumspect in the allocation of blame. Each of the four RUC officers named in Scarva was 'murdered by terrorists', while a memorial in Markethill asserts that those it names 'died as a result of terrorist actions against this province'. A memorial erected in Kilkeel by public subscription in 1999 by a group calling itself 'Mourne Residents for Justice' names eleven local men who were 'murdered by republican terrorists'.

Conclusion

I have shown how conflict commemoration in Northern Ireland has changed through time to incorporate different conflicts. I have also demonstrated how the commemoration of members of the security forces borrows from existing rituals and constructs, but also how it often strikes a slightly different note. The meanings of commemoration are being reshaped through the incorporation of members of the security forces who have died during the Troubles.

Existing war memorials and commemorative ceremonies provide a framework within which to understand death in conflict, and this framework can be used today to give meaning to the deaths of members of the security forces which might otherwise be seen as senseless murder. The inclusion of security force personnel who have died during the Troubles can, however, be seen as contributing to a partial reshaping of that pre-existing commemoration. It is possible to see this reshaping not as something new but instead as part of an ongoing process. The events of the past are always interpreted and reinterpreted in the context of the present, constantly being reshaped over time. As the marking of past events, commemoration is, as Foster (2001, p. 68) has noted, 'always present-minded', and as a result, 'traditions do not only exist in the past. They are actively built in the present also' (Massey, 1995, p. 184).

Public memorials to members of the security forces can be viewed using both of the approaches outlined at the beginning of this chapter. They can be seen as political symbols, and I have discussed how, even in the absence of an agreed national identification amongst Protestants in Northern Ireland, security force memorials could serve to claim existing memorials for the Protestant community. They can also be seen as providing a focus for the public expression of grief, although it should be noted that graves also serve this purpose for the relatives of security force personnel.

The function of memorials as a focus for grief should therefore not be overstated in the case of Troubles commemoration. Given that graves are available, there must be another motivation at work in the perceived need for additional public memorials, and the answer might be found in the quote which begins this chapter. In his speech, Mr Stronge claims that, for the people of Donaghadee, no memorial is needed; their memories are enough. The memorial will, however, serve to remind, 'those passing by ... of the part played by the men of the district in the war'. For all their importance to unionist politicians and the relatives and colleagues of the dead who are often the motivating forces behind their erection, memorials to the security forces also serve this outer-directed purpose, providing a public and permanent reminder of past events which many unionists fear will be intentionally forgotten in the effort to ensure a lasting conclusion to the conflict in Northern Ireland.

References

Barton, B. (1995), *Northern Ireland in the Second World War*, Ulster Historical Foundation, Belfast.

Belfast Newsletter (1926), 'The Donaghadee War Memorial', 2 July, 7.

Bloomfield, K. (1998), *We Will Remember Them: Report of the Northern Ireland Victims Commissioner*, HMSO, Belfast.

Dunn, S. (2000), 'Bloody Sunday and its Commemoration Parades', in T.G. Fraser (ed.), *The Irish Parading Tradition: Following the Drum*, Macmillan, Basingstoke, pp. 129-41.

Foote, K.E. (1997), *Shadowed Ground: America's Landscapes of Violence and Tragedy*, University of Texas Press, Austin, TX.

Foster, R. (2001), 'Remembering 1798', in I. McBride (ed.), *History and Memory in Modern Ireland*, Cambridge University Press, Cambridge, pp. 160-183.

Fraser, T.G. (ed.) (2002), *The Irish Parading Tradition: Following the Drum*, Macmillan Press, Basingstoke.

Fussell, P. (1975), *The Great War and Modern Memory*, Oxford University Press, Oxford.

Gregory, A. and Paseta, S. (eds) (2002), *Ireland and the Great War: A War to Unite Us All?*, Manchester University Press, Manchester.

Inglis, K. S. (1998), *Sacred Places: War Memorials in the Australian Landscape*, Melbourne University Press, Melbourne.

Jeffery, K. (2000), *Ireland and the Great War*, Cambridge University Press, Cambridge.

Johnson, N.C. (2003), *Ireland, the Great War and the Geography of Remembrance*, Cambridge University Press, Cambridge.

King, A. (1998), *Memorials of the Great War in Britain: The Symbolism and Politics of Remembrance*, Berg, Oxford.

Leonard, J. (1997), *Memorials to the Casualties of Conflict: Northern Ireland 1969 to 1997*, Northern Ireland Community Relations Council, Belfast.

Lucy, G. and McClure, E. (eds) (1997), *Remembrance*, Ulster Society, Lurgan.

Longley, M. (1997), 'Say Not Soft Things', in G. Lucy and E. McClure (eds), *Remembrance*, Ulster Society, Lurgan, pp. 121-23.

Longworth, P. (2003), *The Unending Vigil: The History of the Commonwealth War Graves Commission*, Pen and Sword, Barnsley.

Loughlin, J. (2002), 'Mobilising the Sacred Dead: Ulster Unionism, the Great War and the Politics of Remembrance', in A. Gregory and S. Paseta (eds), *Ireland and the Great War: A War to Unite Us All?*, Manchester University Press, Manchester, pp. 133-54.

McBride, I. (ed.) (2001), *History and Memory in Modern Ireland*, Cambridge University Press, Cambridge.

McKittrick, D., Kelters, S., Feeney, B. and Thornton, C. (1999), *Lost Lives: The Stories of the Men, Women and Children Who Died as a Result of the Northern Ireland Troubles*, Mainstream, Edinburgh.

Massey, D. (1995), 'Places and their Pasts', *History Workshop Journal*, 35, pp. 182-92.

Middlebrook, M. (1971), *The First Day of the Somme*, Allen Lane, London.

Mosse, G. L. (1990), *Fallen Soldiers: Reshaping the Memory of the World Wars*, Oxford University Press, Oxford.

National Memorial Arboretum (n.d.), *The National Memorial Arboretum Guide Book*, National Memorial Arboretum, Staffordshire.

Orr, P. (1987), *The Road to the Somme: Men of the Ulster Division Tell Their Story*, Blackstaff, Belfast.

Police Beat (1980), 'L Division Memorial unveiled by CC', **8**(7), p. 35.

Police Beat (1981), 'Memorial to Reservist Unveiled', **3**(6), p. 55.

Police Beat (1982), 'Re-Dedication of Lurgan's Memorial Plaque', **4**(12), p. 15.

Police Beat (1984), 'Unveiling of Memorial Plaque', **6**(6), p. 13.

Ryder, C. (1992), *The UDR: An Instrument of Peace?* Mandarin, London (Revised Edition).

Ryder, C. (2000), *The RUC 1922-2000: A Force Under Fire*, Arrowsmith, London (Fourth Revised Edition).

Whaley, J. (ed.) (1981), *Mirrors of Mortality: Studies in the Social History of Death*, Europa, London.

Winter, J. (1995), *Sites of Memory, Sites of Mourning: The Great War in European Cultural History*, Canto, Cambridge.

Young, J.E. (1993), *The Texture of Memory: Holocaust Memorials and Meaning*, Yale University Press, London and New Haven, CT.

Chapter 10

World Heritage as a Means of Marking Mexican Identity

Bart van der Aa

Introduction

In Mexico City stands a large stone with a text that explains why the Mexicans face so many problems in defining their identity (Groote and Druijven, 2001). On the Plazas de las Tres Culturas, a church built by the Spaniards one can read: 'On August 13, 1521 heroically defended by Cuauhtemoc Tlatelolco fell under the power of Hernan Cortes. It was neither a triumph nor a defeat; it was the painful birth of the Mestizo people that is the Mexico of today'. It is not only striking that a country carves these self-humiliating words in stone at one of its prime monuments; it also shows the divided nature of Mexican identity. This schism in Mexican identity stems from there being three principal cultural groups. They are; the *Indígenas*, the 'original' occupants of the area that nowadays constitutes Mexico (about 15 per cent of the population); descendants of colonists from Spain (10 per cent) and a mixture of these two, the *Mestizo* (75 per cent) (Fischer Weltalmanach, 2001).

One can wonder whether the Mexican identity has always been so divided as indicated on the Plaza de las Tres Culturas. In an attempt to track the recent changes in Mexican identity, this chapter presents an analysis of the changes in the sites that have been nominated for UNESCO world heritage status. The world heritage list is able to show a state's national identity, as the sites nominated by the state parties are designated as 'national flag carriers, symbols in some way of national culture and character' (Shackley, 1998, p. 6).

To track the dominant Mexican identity, one can determine to which population group, and which part of Mexican identity, the selected world heritage sites belong. To this end, the heritage in the Mexican landscape can also be divided into three groups, depending on its age of construction. The pre-colonial heritage is defined here as the objects built before 1492 (by definition the work of the indigenous civilizations); the colonial heritage was constructed between 1492 and

1810; and the post-colonial heritage refers solely to the objects produced after 1810. The year in which a site is built determines which kind of heritage a site belongs to but not what the site represents or means.

Although Mexican monuments can be divided into three time eras, the three population groups remain present in contemporary Mexican society. However, none of the three population groups is expected to be able to determine the content of 'Mexican identity'. Today's ability to mark Mexican identity could depend, among other things, on the size of a population group, the group's voice in the Mexico's political arena, the alleged political correctness of a group's past activities, and the perceived importance of their past. Why are none of the three population groups in Mexico able to fulfil these three pre-requisites?

Mexico's Contested Identities

Indígenas Impotence

The inability of the Indígenas to mark Mexican identity stems from their small share in Mexico's total population, their assumed culturally and economically backward position and the attempt by the Mestizo to transform the Indígenas population into 'real' Mexicans.

First, the number of Indígenas has diminished, both in absolute number and relatively, over the last five hundred years. In absolute numbers, the population has fallen from more than 21 million in 1492 (Evans, 1994) to an estimate of about 13 million in 2001 – constituting today only 14 per cent of the total population. Although Tunbridge (1984) argues that being a minority group does not automatically imply that one cannot be the politically dominant group, this is not the case for the Indígenas in Mexico.

Secondly, the Mestizo who consider themselves as the only 'real' Mexicans look down on the Indígenas as a culturally backward group (Van den Berghe, 1995). Simultaneously, it is difficult for the Indígenas to improve this situation. They are culturally marginal due to a lack of familiarity with Spanish, the formal education system, and the dominant institutions of Mexican culture (Van den Berghe, 1995).

Thirdly, policies have been pursued to promote European, Spanish, and Mestizo culture over the Indígenas to weaken their culture and strengthen that of the Mestizos (Meyer and Sherman, 1991; Morris, 1999). This was necessary as the Indígenas had, once again according to some of the Mestizo population, impeded the unification of the nation. In the last few centuries Mestizos have tried to transform the Indígenas 'both racially and culturally' to accomplish national unity (Morris, 1999). To serve this purpose, the *Instituto Nacional Indígenista* (INI) was created. Although the INI programme, among others, was meant to modernize and empower the Indígenas, according to Van den Berghe, (1995, p. 572), its ultimate purpose was to identify 'the nation with the dominant, Spanish-medium, Mestizo culture.' For Purnell (2001), this process has been very successful; most Indígenas have been transformed into Mexicans. The uprising of the Zapatistas in 1994 was a response to this policy. The uprising should lead to a change in the traditional Mexican (*Mestizo*) views of the Indian and the nation (Morris, 1999), reject the concept of a unified national culture (Cook and Joo, 2001) in favour of improving the economic and social position of the Indígenas.

The cultural, economic and political position of the Indígenas within Mexican society does not lead to any expectation that the descendants of the former Toltec, Mixtec, Zapotec, Aztec and Maya civilizations have had many possibilities to establish a Mexican identity that is based on the pre-colonial past. This is in spite of the presence of highly valued complexes in the Mexican landscape that were built by the indigenous civilizations, such as Teotihuacan and Monte Alban.

Colonial Terror

The Spanish colonists were politically superior after Columbus' 'discovery' of America in 1492. However, the position of the Spanish in Mexican society has eroded over the centuries that followed. The fight for independence from Spain in 1810 resulted partly from the recognition of the injustice of the conquest (Brading, 2001). Furthermore, at the time of the Mexican Revolution (1910-1917), the Spanish contributions to the hybrid culture were often de-emphasized or even denied and denigrated (Van den Berghe, 1995).

The negatively associated values of the Spanish Conquest is not only a burden for the Spaniards, but also for the largest part of the Mexican population, the Mestizo, who are haunted by the anything but peaceful colonization of the Americas by the Spaniards. How can the Mestizos – being a mixture of Indígenas and Spanish blood – identify themselves in any reasonable way with their Spanish part when they are aware that one part murdered the other? This internal dilemma that hardly any Mestizo can escape from is strengthened by Lynch's (1972) observation, that a desirable image is one that is in tune with ones own biological nature. This complexity can also be found in the built environment. The colonists repeatedly built their Catholic churches on top of, and often with the stones from, demolished holy pre-colonial complexes.

As such, the prevailing thought today about colonial Mexico is one that sees the Spaniards as 'culturally backward'. Even though the Colonists did build some splendid colonial towns, it is not expected that this era of Mexican history will be chosen in order to establish a Mexican identity.

Young Nation, 'Blank' History

Already at the time of independence from Spain in 1810, Mexican society was in search of 'who they were'. According to the nationalists, 'only the Mestizos were true Mexicans, since creole [that is, offspring of Spanish colonists] landowners were European in cultural affiliation and Indians were bound by the parochial loyalties of their *pueblos* [people]...' (Brading, 2001, p. 528). Having made the choice of rejecting all that is Indigenous and Spanish, the dominant group lumbered itself with an almost impossible task. Turning themselves away from both the pre-colonial and colonial eras meant at the same time that the Mexican state would have no historical roots to legitimate its rule. From then on, Mexico lacked an uncontested culture to shape its identity.

This search for a Mexican identity continues until today: 'Now, Mexico is fulfilling the complexity of its Mestizo legacy. Mexicans are realizing there is nothing pure here... Mexico's skin is white, black and a dozen shades of brown. And so is its heart' (Martinez, 1997). Nonetheless, the last two hundred years have yielded some 'cultural production'. Worth mentioning in particular are the murals painted by Rivera and Orozco. These murals are wall paintings on public buildings that show important scenes from Mexico's history. However, the originality of these paintings as pure Mestizo could be questioned as these modern Mexican artists often found inspiration for their paintings in the pre-colonial era (Brading, 2001). Furthermore, the Mestizo era has not yet seen a Golden Age that can compete with the cultural production, in terms of building outstanding cities, as during the heydays of the Aztecs, Maya's, or Spaniards.

The weaknesses of the three major ethnic groups in Mexico support the expectation that Mexico will face problems in the process of marking its identity. However, why is it important for Mexico to have its own identity in the first place?

Identity Marked in Heritage

National identity can be a centripetal force for any state in strengthening its coherence. This identity, which is often anchored on myths, legends, and events from the past, should support the idea that the inhabitants of a state have the common feeling that they live within, and are part of, an imagined community (Anderson, 1983). This construction process is not only based on the intangible past as myths and legends; also tangible relics of the past can play a major role in this process of nation building. As Lowenthal (1993) has pointed out landscape has become a compelling symbol of national identity. More in general, heritage, defined as the contemporary uses of the past (Graham et al., 2000), is often used for this purpose.

Ashworth and Larkham (1994, p. 7) correctly add to this definition that heritage not only 'has a proven track record of outstanding success in formulating and reinforcing place-identities,' but also that this is 'in support of *particular state entities*' (my italics). This emphasis is important, as '... identities are never unified and, in late modern times, increasingly fragmented and fractured; never singular but multiple' (Hall, 1996, p. 3). In other words, the selected heritage only shows the identity that 'belongs' to the community that selected it. Will the heritage of the unrepresented or underrepresented population groups also be selected?

This question is most prominent in those states that consist of two or more ethnic groups. It is often in these societies that the relics of ethnic minorities face an uncertain future, more often being disguised than glorified. This concealment stems from a state's awareness that unity should be rooted in a consistent cultural background (Coremans, 1979). At the same time, however, such a one-sided approach towards history is unwelcome, as it threatens the position of minorities (Halonen, 2001).

The selection of heritage can become a complicated and painful process in states that have been colonized. The 'original' population in these states has often been violently dominated by its colonizers and the present population faces the difficult task of how to deal with this contested past. This complexity reveals itself partly in the kinds of world heritage sites that the Latin American countries have chosen to nominate and the size of each population group. The population in most Latin American countries can be divided into at least three major ethnic groups; indigenous, white colonists and a mix of these two groups – sometimes with the addition of the descendants of the former black slaves from Africa.

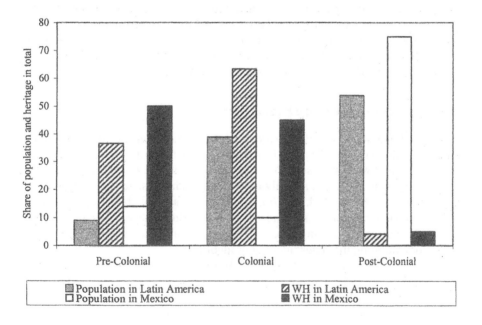

Figure 10.1 Share of population groups and their heritage in Latin America and Mexico

Source: Fischer Weltalmanach (2001) and UNESCO (2002)

The majority of Latin Americans are Mestizo. However, only a very small number of world heritage sites can be associated with this population group (Figure 10.1). At the same time, the number of sites that can be associated with the indigenous and white population groups is much larger than seems to be justified by their present population sizes. The question here is whether this imbalance is related to the problems of Latin American states in establishing their national identity, as Graham et al. (2000) argue.

A study of Mexico could be seen as exemplifying the situation in Latin American countries as a result of Mexico's contested past as demonstrated by the Plaza de las Tres Culturas. Figure 10.1 shows the world heritage sites nominated by the Mexican government. Given the size of the three population groups in Mexico, the share of heritage sites from the pre-colonial and colonial era is larger than the shares in population would justify. The opposite is true for the sites from the post-colonial era. As such, it seems that the Mexicans have only been able to nominate heritage sites by representing Mexico partially, selective and in a distorted manner (Conkin and Stromberg, 1989).

It is striking that Mexico has been able to nominate sites, as this was thought impossible. Two important reasons for nominating world heritage sites are the possibility of attracting tourists to these sites and to mark Mexican identity. This study only takes into consideration the importance of heritage in the process of establishing a common Mexican identity. What is it exactly that Mexico has nominated for world heritage status and how has this changed over time?

Dynamics of World Heritage Sites in Mexico

World heritage status is only rewarded to those sites that are of 'outstanding universal value' (UNESCO, 2002). It is in the first place this criterion that states have to keep in mind when they select possible sites for listing. At the same time, the world heritage list is often primarily used to aggrandize a state's past (Tunbridge and Ashworth, 1996; Pocock, 1997a). Added to the fact that about eighty per cent of all nominated cultural sites do indeed receive world heritage status, one can use the world heritage list, being a list of 'national icons' (Lowenthal, 1998), as a means of exhibiting the states' identity or identities.

Mexico, which ratified the 1972 World Heritage Convention in 1984, received its first world heritage site in 1987. It has been quite successful since then, with twenty cultural and two natural designated sites. With 22 world heritage sites, Mexico occupies the ninth place in the world. The total number of cultural sites in Mexico is increased by two in this chapter. Two world heritage sites in Mexico have been split, as they are a combination of colonial and pre-colonial heritage (Mexico City combined with Xochimilco and the city of Oaxaca with Monte Alban). Furthermore, no cultural and two natural sites (nature reserve El Triunfo and Patzcuaro Lake) have been rejected by UNESCO.

An analysis of what is most important in the course of Mexican history, according to what this country has nominated in the past two decades and may nominate in the near future for the world heritage status, yields a clear picture. First, regarding what already has been listed, the pre-colonial and colonial era are represented equally in absolute numbers. Secondly, concerning what Mexico is planning to nominate in the future from its tentative list that mentions 22 sites in the Autumn of 2002, a fast increasing awareness of the relevance of sites from the post-colonial era can be observed.

The list of nominated sites by the responsible agency for cultural world heritage sites in Mexico, the *Instituto Nacional de Antropología e Historia* (INAH), strongly suggests that this agency has done its utmost to keep the number of pre-colonial and colonial sites in balance throughout the last two decades (Figure 10.2). One by one, sites from the pre-colonial and colonial era have been submitted to

Year of nomination	Pre-Columbian	Colonial	Post-Colonial
2002	Calakmul		
1999	Xochicalco	Campeche	
1998	Paquimé	Tlacotalpan	
1997			Hospicio Cabañas
1996	Uxmal	Querétaro	
1994		Popocatepetl	
1993	Sierra San Francico		
1992		Zacatecas	
1991	El Tajin	Morelia	
1988	Chichen Itza	Guanuajato	
	Palenque		
1987	Teotihuacan	Puebla	
	Monte Alban	Mexico City	
	Xochimilco	Oaxaca	

Figure 10.2 Listed world heritage sites in Mexico, categorized into three periods

Source: UNESCO (2002), adapted data

UNESCO. Consequently, this has led to a significant number of sites from before 1810.

The numerous pre-colonial and colonial sites are accompanied by the almost complete absence of post-colonial sites until now. *Hospicio Cabañas* in Guadalajara is the only site that belongs to this category. This former centre for the disabled is especially known for its murals, painted by Orozco in the late 1930s, depicting the violence of both the pre-Columbian world and the Spanish Conquest (Meyer and Sherman, 1991).

However, in the near future, this imbalance between pre-colonial and colonial heritage on the one hand and post-colonial heritage on the other may become less. This expectation is based on the changed character of the twenty-two properties on the last tentative list that Mexico submitted to UNESCO in 2002 (Figure 10.3). This list is a prime example of the temporal character of any heritage selection. In Mexico, its focus has shifted from the pre-colonial era (from eleven to five sites) to the post-colonial one (from one to seven sites), whereas the attention to the colonial era has remained the same (ten sites).

Pre-Columbian	Colonial	Post-Colonial
Mitla	Historic Town of Alamos	Ciudad Universitaria
Pre-Hispanic City of Cantona	San Luis Potosí	Rivera and Kahlo's museum
Chicomostoc-La Quemada	Franciscan missions	Mies v/d Rohe and Candela's
Ahuehuete Tree of Tule	San Sebastián del Oeste	Luis Barragan's home museum
Caves of Yagul and Mitla	Churches in the Zoque	Monterrey's Industrial Facilities
	Santa Prisca	Railway Station Aguascalientes
	Camino Real	Chapultepec woods and castle
	Aqueduct of Padre Tembleque	
	Agave Landscape	
	Jesuit colleges in Tepotzotlán	

Figure 10.3 Sites on Mexico's tentative world heritage list, categorized into three periods

Source: INAH (2002) adapted data

Especially striking is the strongly increased interest in twentieth-century relics, including, among others, the *home study museum* of Diego Rivera and Frida Kahlo, Luis Barragan's home museum, and *Ciudad Universitaria*. To underline this shift, Mexico will only put forward the first two mentioned sites in 2003. The attention to these post-colonial sites should not only be seen as an orientation *towards* new kinds of heritage that might represent 'real' Mexicanness. It is, at the same time, a movement *away from* the former, one-sided, stress on pre-colonial and colonial heritage. For example, Frida Kahlo's (1907-1954) work is steeped in Mexicanness with her Mexican identity being an important element in her painting. This is dramatically illustrated in *My nurse and I* (1937), in which the artist, here depicted as an infant with an adult's face, is nursed by an Indian woman whose massive, expressionless form is reminiscent of pre-Columbian sculpture (Meyer and Sherman, 1991). Furthermore, the building of Ciudad Universitaria from 1954 not only shows one of the last murals in Mexico that can be found in public space. The functional, modern building also dissociates itself from the former university building in Mexico City, which exhaled a decorative colonial splendour (Ibid.). How could this shift in emphasis be explained?

Mexican World Heritage Sites in a Global Perspective

This selection of heritage sites in Mexico is, as usual, highly influenced by who has selected it. Generally it is claimed that the selection depends on the cultural background of the people who select the sites (Pocock, 1997b). In Mexico, the outcome of the selection of world heritage sites is also influenced by the organizational structure of INAH. This is strengthened by UNESCO's policy on the nomination of world heritage sites. Only nominations from organizations that are officially appointed by the state party are allowed. Nominations from other organizations, such as non-governmental organizations or governments at a lower scale-level, will not be dealt with.

   In the 1980s, INAH's structure was in accordance with the 'organic law', which divided the department into two sections: the *office of pre-Hispanic monuments*, and the *office of colonial monuments* (INAH, 2002). The subdivision of this state organization mirrors exactly the kind of heritage that has been nominated for world heritage year by year. Consequently, one cannot escape the thought that the dual structure has contributed largely to what has been selected by INAH. Furthermore, one wonders why INAH was asked to select Mexico's potential world heritage sites, as the outcome could be predicted beforehand, namely, a neat balance of pre-colonial and colonial heritage and no nominations from Mexico's post-colonial era.

   In the course of the 1990s, INAH's internal structure changed. The organization came under the co-ordination of the *Consejo Nacional para la Cultura y las Artes* (National Council for Culture and the Arts), which allowed a shift away from the focus on the pre-colonial and colonial era (INAH, 2002). From then on, INAH's structure was divided into three fields; archaeological, historical, and artistic. Again, this structure is clearly reflected in the composition of Mexico's last tentative list. Apparently, Mexico's national identity and its dynamics are reflected in the organizational structure of INAH, which immediately influences the outcome of the selection of world heritage sites.

The selection of world heritage sites is, like any selection, by definition open to discussion as selection always includes exclusion. Some authors have questioned why some Mexican sites have not been nominated for the world heritage status. For example, Evans, 2002, states that 'other Mayan sites, including the oldest known of Dzibilchaltun, Mayapan, dot the province [of Yucatan], but are not world heritage sites, an indication of both the arbitrary nature of inscription and limited conservation and site management resources'. Furthermore, Brading (2001) emphasizes that the place where the Virgin Mary appeared in 1531 is most Mexican compared even to the political and cultural symbolism of Teotihuacan. The Virgin of Guadalupe would have been especially expected to be on Mexico's tentative list of 2002, as the world heritage committee has enlarged the possibilities for listing sites that are not tangible, such as sites that are renowned for legends and myths that are located at a place (Pocock, 1997). Similarly, one could question whether particular sites do deserve a world heritage listing as it is not clear how they meet the criterion of outstanding universal value.

Instead of focussing on whether *particular* sites do deserve listing or not, one can also discuss whether the present representation of the three distinguished *types* of heritage is justified. There seems to be fertile ground for such a discussion in Mexico, as the number of *national* monuments from the pre-colonial period outnumbers the ones from the colonial era (INAH, 2002). There are about 200,000 sites from the era before the arrival of Columbus in America against 110,000 monuments from between the sixteenth and nineteenth centuries. Furthermore, the pre-colonial era lasted much longer than the colonial one. Is INAH's approach of keeping the two oldest categories of heritage in balance on the world heritage list a 'fair' one? A comparison with the global scale-level offers one way of looking at this question. The distribution of world heritage sites in Mexico, according to the year in which the sites were built, fits fairly well within the distribution at the global scale-level. A comparison between the two distributions yields two points worth mentioning. First, there are relatively many sites from Mexico's colonial era. This deviation disappears partly in a global context, as the share of sites from the sixteenth century is also largest at the global level. This global trend can be explained by a combination of factors, including the increasing possibility of sites surviving as they are younger and the necessity to be of some 'age' to become exceptional enough for listing. This similarity on both scale-levels is reflected in the average age of cultural world heritage sites. In Mexico, the average age of listed cultural world heritage sites is 860 years, whereas the average age at the global scale-level is about one thousand years old.

A second striking element in the distribution of the age of Mexican sites in comparison with the global scale is the relatively large number of sites from the twentieth century on the tentative list. This shift is reflected in the fall of the average age of the cultural sites on the tentative list to 'only' 360 years. This leads to the observation that Mexico has many more sites from the twentieth century than the rest of the world. Mexico's shift towards younger sites indicates that Mexico still has to mark its identity. Mexico's identity can neither be based on the pre-colonial, nor on the colonial period. Consequently, it should be found in the post-colonial period to honour the work of the Mestizo.

There is, however, another explanation that weakens Mexico's striking shift towards younger heritage. Mexico's attention to the twentieth century manifests itself only on its tentative list, whereas the tentative lists of the rest of the world are not easy to calculate. Nonetheless, there are two indications that the average age of nominated sites is also decreasing in the rest of the world. First, at the global scale, the average age of listed cultural sites has decreased from roughly 1200 years old in 1978 to 800 years old in 2002. This is partly the impact of UNESCO's request for the nomination of those kinds of sites that are underrepresented on the present world heritage list, as stated in the Global Strategy (UNESCO, 1994). Secondly, a comparison with the tentative list of another active state, the United Kingdom (UK), reveals that this state also puts more stress on younger heritage in comparison with the sites on its latest tentative list from 1999. Mexico puts most emphasis on the twentieth century, whereas the UK stresses the nineteenth century due to its potential industrial heritage (DCMS, 1999).

Nonetheless, the large number of sites in Mexico from the sixteenth century cannot entirely be explained by the distribution at the global scale, as the number of sites from the sixteenth century is more than three times higher in Mexico than in the rest of the world. However, one should be conscious that a temporal distribution of sites at the global scale is always smoother than in a single state. States tend to have an economic and cultural peak in a certain period, whereas the arithmetical average of the peaks of all countries in the world is more evenly distributed over time. Furthermore, the share of listed sites from the three centuries that make up the colonial era in Mexico is only 1.5 times higher than at the global scale (Table 10.1).

Table 10.1   Share of listed world heritage sites from three eras at two scale levels (per cent)

	Mexico	World
Pre-colonial (until fifteenth century)	50	61
Colonial (sixteenth until eighteenth century)	45	30
Post-colonial (nineteenth and twentieth century)	5	9
$N =$	22	*564*

Source: UNESCO (2002), adapted data

All in all, it appears that the temporal distribution of Mexican world heritage sites is only slightly different from the international distribution. Possibly the share of colonial heritage might be a little above the number of sites from that period at the global scale. How does the selection of world heritage sites in Mexico relate to what is generally regarded as 'Mexican identity'?

Mexican National Identity Reflected in Heritage Selection?

According to Renes (1999), the national identity of any state is anchored in two events that characterize the nation. These are the myth surrounding the origin of the state and its economic and cultural heydays, the 'Golden Age'. *The* event in the course of Mexican history that marks out its origin is the appearance of the Virgin of Guadalupe before the baptised Indian Juan Diego. Mexico's Golden Age can, arguably, be found in the pre-colonial period. These are the days of the Aztec and Mayan civilizations, when they constructed the highly valued complexes such as Teotihuacan, Chichen Itza and Palenque.

These two anchors are also mentioned by Morris (1999, p. 364) when he describes Mexican identity:

> Mexican national identity contains a wide mix of expressions and components. It encapsulates ... (a) *Mestizaje,* with the *mestizo* representing the racial expression of the nation, and as an example of... a superior mixture and the future saviour of the world; (b) pride in past Indian civilisations once occupying the territory, including a sense of 'primordial disintegration' or the idea of a great and glorious history interrupted, diverted and degraded; and (c) a reverence for the Virgin of Guadalupe, seen generally as God's confirmation of the nation's spiritual and even racial uniqueness... Both the intensity and content of Mexican national identity are also rooted in, and shaped by, juxtaposition to the USA: the predominant 'other'... being Mexican means not being *gringo*.

According to Morris, Mexican identity is also built upon another pillar – *Mestizaje* – and is sharpened by the overall emphasis that being Mexican is anything but being related to the United States of America. The third principle is probably the outcome of the Mestizos awareness that only '... a remembered past becomes a crucial part of our self-image' (Conkin and Stromberg, 1989). The latter recognition will have played a role in Mexico's cautious, but resolute move towards the appreciation of the post-colonial era in the selection of world heritage sites.

Furthermore, the construction of one's own identity by stressing the differences with others can be successful, as 'identities are constructed through, not outside, difference' (Hall, 1996). In this respect, heritage from all Mexican eras could be used to distinguish Mexicans from all other North Americans in terms of ethnicity and religion (Martinez, 1997). Based on this general image of how the Mestizo see themselves, the selection of most world heritage sites is not surprising. The murals represent what is 'real' Mestizo, pre-colonial complexes confirm Mexico's roots, and the Catholic colonial cities stress the difference with the United States of America. Nonetheless, there are two issues worth mentioning at this point. First, there is no attention to the Indígenas of today, and secondly, the interest for the colonial era seems exaggerated.

By selecting the heritage of the past that lies far behind us, Mexico tries to represent itself as a state with long and strong established roots. However, this presumed continuity in the course of Mexican history is in fact a discontinuity. The Indígenas who are glorified on the world heritage list are not the present-day Indígenas (Brading, 2001), but the ones that died out five hundred years ago. The Indígenas of today differ in many aspects from the 'original' indigenous population. The civilizations of the past used to live in urban centres, whereas the Indígenas of today live in the countryside. Their low position in Mexican society after 1492 is reflected in the total absence on the world heritage list of sites that have been built by the Indígenas after the early sixteenth century. In simple terms, the Indígenas from after the fifteenth century are excluded. This also explains why there can be pre-colonial heritage from Mexico on the world heritage list. The sites are selected by INAH to mark Mexico's long roots; not to represent the Indígenas of today. The Mexican government must also have thought in such ambiguous terms about the Indígenas in the past, as appears from INAH's objectives when it was established in the late 1930s. INAH had to improve the preservation of *national monuments* and the study of *indigenous groups* (INAH, 2002).

Secondly, there is no basis in Mexican identity for selecting the large number of sites from the colonial era. It appears that Mexican identity is largely based on the pre-colonial and post-colonial past and that hardly any reference is made to the colonial era. The dominant role of the pre-colonial era over the colonial one is illustrated by the destruction of colonial buildings after the 're-discovery' of an Aztec temple in the former city of Tenochtitlán, present Mexico City. The glorification of the former Aztec empire was necessary as it had to become one of the pillars on which the Mestizo society could build its new roots (Hamnett, 1999). However, this representation is in conflict with how Mexico in general pictures itself. In this perspective, the large number of world heritage sites from the colonial era must be seen as an exaggeration.

Conclusion

The selected world heritage sites in Mexico do not entirely reflect what generally is assumed to be 'Mexican identity'. Especially the large number of world heritage sites from the colonial era cannot be explained by the idea of a Mexican identity. This may well be explained by other factors that influence the selection of world heritage sites that have not been included in this chapter, such as the potential to attract more tourists to world heritage sites.

Nonetheless, this chapter has shown that the change in what is selected as a world heritage site reflects changes in Mexican identity. The neat balance of nominations from the pre-colonial and colonial era for the world heritage list up to the mid 1990s reflects Mexico's struggle over its choice of Indígenas or Spanish identity. Furthermore, the shift towards post-colonial heritage in the late 1990s shows the first signs of an awakening country that is becoming aware of its post-colonial achievements. It appears that the selection of heritage, and thus what is regarded as the national identity, can change significantly in less than twenty years. World heritage can help to create such a new identity. It should be noted that what is 'Mexican identity' in this respect is defined by the policymakers of INAH.

The present selection of Mexican world heritage sites not only reflects Mexican identity. Certain parts of Mexican history seem to be exaggerated or completely ignored. The weak position of Spanish identity in present-day Mexico seems not to justify the large number of world heritage sites in Mexico from the colonial era. Apparently, the relevance of Mexico's colonial heritage is not well understood. Either the importance of Mexico's colonial era for marking its identity is larger than generally assumed or the colonial cities are listed for other reasons than marking Mexican identity.

In the latter case, the search for a 'balanced representation' of Mexican identity might be an impossible one. Potential world heritage sites are primarily selected for their 'outstanding universal value' (UNESCO, 2002), not to mark a certain national identity of a state. However, it is not clear who should classify the site's quality as 'outstanding' and which characteristics make a site important – its historic importance, its state of preservation, the share of the current population?

The expectation was that the heritage of all eras would face difficulties in being selected as world heritage sites. However, each era will soon have a rather substantial position on the world heritage list. This does not, however, assure that all population groups are evenly represented.

References

Anderson, B. (1983), *Imagined Communities Reflections on the Origin and Spread of Nationalism*, Verso, London.

Ashworth, G.J. and Larkham, P. J. (1994), 'A Heritage for Europe: The Need, the Task, and the Contribution', in G.J. Ashworth and P. Larkham (eds) *Building a New Heritage: Tourism, Culture and Identity in the New Europe*, Routledge, London, pp. 1-15.

Berghe, P. L. van den (1995), 'Marketing Mayas: Ethnic Tourism Promotion in Mexico', *Annals of Tourism Research*, 22(3), pp. 568-88.

Brading, D.A. (2001), 'Monuments and Nationalism in Modern Mexico', *Nations and Nationalism*, 7(4), pp. 521-31.

Conkin, P. K. and Stromberg, R.N. (1989), *Heritage and Challenge: The History and Theory of History*, Forum Press, Illinois.

Cook, S. and Joo, J-T. (2001), 'Ethnicity and Economy in Rural Mexico: A Critique of the Indigenista Approach', *Latin American Research Review*, 30(2), pp. 33-59.

Coremans, P. (1979), *Organization of a National Service for the Preservation of Cultural Property*, in UNESCO (ed.), *The Conservation of Cultural Property: With Special Reference to Tropical Conditions*, Les Presses de Gedit, Tournai.

DCMS (1999), *World Heritage Sites: The tentative list of the United Kingdom of Great Britain and Northern Ireland*, DCMS, London.

Evans, G. (1994), *Whose Culture is it Anyway? Tourism in Greater Mexico and the Indígena*, in A.V. Seaton, (ed.), *Tourism: The State of the Art*, John Wiley, Chichester, pp. 131-47.

Evans, G. (2002), 'Mundo Maya - World Heritage in Post-*Colonial* Mesoamerica', paper presented at the IICTD conference, University of North London, 2-4 September.

Fischer Weltalmanach (2000), *Der Fischer Weltalmanach: Zahlen, Daten, Fakten*, Fischer Taschenbuch Verlag, Frankfurt am Main.

Graham, B., Ashworth, G.J. and Tunbridge, J.E. (2000), *A Geography of Heritage: Power, Culture and Economy*, Arnold, London.

Groote, P. and Druijven, P. (2001), 'De Zapatista's *Postmoderne Rebellen*', *Geografie*, **10**(4), pp. 29-32.

Hall, S. (1996), 'Introduction: Who Needs "Identity"?', in S. Hall and P. du Gay (eds), *Questions of Cultural Identity*, Sage, London, pp. 1-7.

Halonen, T. (2001), 'World Heritage Committee', *Presidents and Prime Ministers*, **10**(6), pp. 19-34.

Hamnett, B. (1999), *A Concise History of Mexico*, Cambridge University Press, Cambridge.

Hooff, H. van (1995), 'Bewahrung des Weltkulturerbes in Lateinamerika und der Karibik', *Geographische Rundschau*, **47**(6), pp. 355-59.

INAH (2002), *El Patrimonio de México y su Valor Universal: Lista Indicativa*, INAH Mexico, D.F.

Johnston, R.H., Gregory, D., Pratt, G. and Watts, M. (2000), *The Dictionary of Human Geography*, Blackwell, Oxford.

Lowenthal, D. (1993), 'Landscape as Heritage: National Scenes and Global Changes', in J.M. Fladmark, (ed.) *Heritage: Conservation, Interpretation and Enterprise*, Donhead, London.

Lowenthal, D. (1998), *The Heritage Crusade and the Spoils of History*, Cambridge University Press, Cambridge.

Lynch, K. (1972), *What Time is this Place?*, MIT Press, Cambridge.

Martinez, R. (1997), 'Mexico's Search for Itself', *Nation*, **264**(16), pp. 22-4.

Meyer, M.C. and Sherman, W.L. (1991), *The Course of Mexican History*, Oxford University Press, New York.

Morris, S.D. (1999), 'Reforming the Nation: Mexican Nationalism in Context', *Journal of Latin American Studies*, **31**(2), pp. 363-97.

Pocock, D. (1997a), 'Some Reflections on World Heritage', *Area*, **29**(3), pp. 260-68.

Pocock, D. (1997b), 'The UK World Heritage', *Geography*, **82**, pp. 380-85.

Purnell, J. (2002), 'Citizens and Sons of the *Pueblo*: National and Local Identities in the Making of the Mexican Nation', *Ethnic and Racial Studies*, **25**(2), pp. 213-37.

Renes, H. (1999), 'Landschap en Regionale Identiteit', *Geografie*, **8**(2), pp. 8-15.

Shackley, M. (1998), 'Introduction: World Cultural Heritage Sites', in M. Shackley (ed.), *Visitor Management: Case Studies from World Heritage Sites*, Butterworth Heinemann, Oxford, pp. 1-14.

Tunbridge, J.E. (1984), 'Whose Heritage to Conserve? Cross-cultural Reflections on Political Dominance and Urban Heritage Conservation', *Canadian Geographer*, **28**(2), pp. 171-80.

Tunbridge, J.E. and Ashworth, G.J. (1996), *Dissonant Heritage: The Management of the Past as a Resource in Conflict*, Wiley, Chichester.

UNESCO (1994), 'Expert Meeting on the 'Global Strategy' and Thematic Studies for a Representative World Heritage List', Paris: 20-22 June.

THEME III:
INSIDERS AND OUTSIDERS

Introduction to Theme Three

The Editors

Part Two focussed upon the creation of places by official agencies acting in some public interest. Here, we shift the viewpoint to the consumer, largely as citizen and local resident. This raises the question: 'are popular senses of place significantly different from official ones or can the latter only ever be an aggregate and expression of the former?' Again, do the two coexist at different scales and with different purposes? The imposition of heritage designations alters the character of a building or area, which thus becomes monumental and historic with potential consequences for the sense of place held by insiders and outsiders. If there is a collective sense of place, then what are the place identities of those who – for whatever reasons– feel themselves to be ignored, alienated and marginalized? They may be displaced by such exclusion in a cultural, social or even physical sense but they may also attempt to create places for themselves which stand outside official representations. In a real sense, in these territories, the outsiders become insiders.

Van Hoven, Meijering and Huigen [11] concentrate upon the relationship between social alienation and physical displacement. They discuss how 'drop-outs' from mainstream society, feeling constrained by being 'out of place' at one location, physically displace themselves in order to relocate somewhere with a sense of place that is more congenial to their way of life, or is capable of being so moulded. Such displacement is a familiar pattern for religious minorities whose practices and way of life required self-government and separation from others. This case, however, concerns groups seeking more ideological and artistic freedom than their current place can provide. The chronological sequence they describe is first a growing feeling of being 'out of place', which prompts a voluntary migration, a 're-placement' in search of a place less constraining and more conducive to the group's ideas and pattern of life. This is not the end of the story, as over time the cohesion of the group's identity may weaken or the initially hostile society may change leading to a re-incorporation into the mainstream. The external environment that triggered the reaction is itself not static and immutable. In this case, the ideological background itself changed from Weimar republic, National Socialism, Stalinist centralism, post-Stalinist communism, to the current liberal federal republic. What was liberation from a constraining or alien sense of place becomes a lost and re-found identity.

An initial axiom of this book is that places are the product of creative imaginations. Artists, and particularly writers, have long created imagined places with no reference to actual locations. Many however have also set their work in an existing place and interweave the settings and the narratives as a 'landscape with figures' or as 'figures in a landscape'. The novels analysed by Stainer [12] are describing different senses of place, the different 'Belfasts' of the authors, even if located in the same physical space of the city. The place becomes more than just a stage for the characters and events described and in turn the story shapes the place in the imagination of the reader. Stainer argues that, within these depictions, we can see portrayed a more inclusive and 'progressive' conceptualization of social space, descriptively and theoretically one which perhaps questions boundaries and categories.

The Newfoundland of the imagination [13] was a composite created over the last 100 years, in part by outsiders idealizing the prosaic practices of the fishing way of life, and in part as deliberate promotion intended mostly for consumption elsewhere. Over time, the inhabitants adopted the externally projected image of themselves. Eventually an image based loosely upon the economic, social and settlement realities of the fishing 'outport', survived quite independently of such realities. The quite abrupt termination of the fishing based society cut the sense of place adrift from any deterministic origins. The image, far from withering as an anachronism, is robustly extending and strengthening its hold on imaginations inside and outside the 'Rock', trapping in the process the people in their own sense of time.

Groote and Haartsen [14] explore the development of representations of rural areas over time by analysing the contents of a long running television programme on rural areas in the Netherlands. The popular senses of the rural not only vary between different actors but also through time as meanings and values are continuously produced and reproduced in the representation of such places. By not confining their attention to describing the past, the authors cautiously raise a tantalizing prospect that if the mechanisms could be understood then there is a possibility for at least forecasting, if not influencing, future representations.

The final chapter shifts the focus from rural to urban and narrows the scale to a single neighbourhood. Kuipers [15] examines the reactions of local residents to a conservational designation imposed from above. The Korreweg district of Groningen has recently become an urban conservation area as a result of a valuation of its historic and aesthetic qualities by national and local authorities. Local residents have long endowed the district with a distinct identity but one that was based on quite different characteristics than those now recognized by 'outsiders'. It is likely that local personal histories and familiarities, frequently not evident to outsiders, play a larger role in the shaping of local senses of place than wider historic associations, architecture and visual forms in general. The consequences of this process are that two senses of place exist which may conflict with each other, become mutually reinforcing through convergence or just coexist in mutual ignorance or indifference. In any event, the argument over externally imposed and locally evolved identities, and the roles of insiders and outsiders, returns the discussion to the introductory model with which this book began.

Chapter 11

Escaping Times and Places: An Artist Community in Germany

Bettina van Hoven, Louise Meijering and Paulus P.P. Huigen

Introduction

Within the study of cultural geography, considerable attention has been given to involuntarily excluded 'Others' (Cloke and Little, 1997), such as the poor (Duncan and Lamborghini, 1994), the homeless (Cloke et al., 2001; Dear and Wolch, 1987), the elderly (Russel and Schofield, 1999), the disabled and ethnic minorities (Samers, 1998). In certain places, these groups are considered 'Other', that is, not belonging and out of place, while dominant groups marginalize these 'minorities' in order to preserve the characteristics and qualities of 'their' place.

A significant group of 'Others', however, comprise those who *voluntarily* withdraw from urban society and these have received considerably less attention in the literature. These people commonly share their disapproval of the dominant norms and values in the city and reject the place itself. Urban-rural migration illustrates the spatial consequences of discomfort experienced in the city and the need to move to better, more quiet or safer places. There are, however, examples of people who have focused and structured their collective disgust with mainstream society and formed communities elsewhere, often in rural areas as well. An interesting example of such communities is one of artists on the island of Usedom in the former German Democratic Republic (GDR). The origin of this community can be traced back to the early rise of Fascism prior to World War II when a number of less compliant members of society established their own way of life in a remote place, largely untouched by this ideology. However, collective escapism to promising places does not necessarily guarantee contentment without an expiration date. Changing times influence life outside and in these new communities, thus challenging people's histories, homes and identities all over again. After a short discussion of the relationship between identity, place and 'intentional communities', the focus of this chapter is on the relationship to place of a community of artists in Germany.

Identity and Place

At the heart of nonconformist movements away from mainstream society and urban settings are questions of identity which can be defined as 'the story we tell of ourselves and which is also the story others tell of us' (Robertson et al., 1994, p. 95). As these two stories may significantly differ from each other, the concept of identity remains rather convoluted. Identity changes continuously: it is fluid (Hatty, 1996) and largely developed through the process of 'othering', which means that an identity is not positively defined in terms of what it consists of, but negatively in terms of what it is *not*, (Minh-ha, 1994). Many different identities – individual, group and place – can be distinguished. The relation between place and identity is important (Teather, 1999; McHugh, 2000; Mitchell, 2000) and changes in place usually impact upon identity formation at least to some degree. Rutherford (1990, p. 24) argues that a feeling of 'not belonging', a negative sense of place, or a feeling of displacement (although he does not use this term) is endemic in our society. He perceives a 'sense of unreality, isolation and being fundamentally 'out of touch' with the world' as a result of a confusing multitude of potential identities that are ascribed to us. Where opportunities arise, feeling out of place can result in migration to a different place that may be more accommodating toward one's identity. The movement expresses hope for a positive sense of place, in which a new home and identity can be created over time (see also Robertson et al., 1994). Increasing attachment to the new place can consolidate desired identities and/or change them. But perceptions of places are never static either and changes may occur that are again at odds with who and where one wants to be at a particular time in one's life. However, even without a physical move, individual identity can be challenged by changes within one's place. If a physical relocation is not desirable or possible, a person may retreat to places from times that already passed, a world that merely exists within, a kind of 'internal migration'. In the case of the unified Germany, many East Germans remained attached to their old lives under socialism and found it difficult to adjust and feel at home in the New Germany. This experience has also been termed 'Ostalgia', combining nostalgia with the affection felt for the former GDR (see also Hörschelmann and van Hoven, 2003).

Place, Identity and 'Intentional Communities'

Perry et al. (1986) argue that the phenomenon of 'urban dropout' was symptomatic of the 1960s and 1970s and was limited largely to the hippy communes. In these cases, the peace movement and the protests against conventional 'stiffdom' of the (American) post-war society were manifested in attempts to create a better, 'truer' way of life by a group of people who identified as social dissidents. Most of these communities were short-lived. Hetherington (1998) revisited the phenomenon of 'urban dropouts' and defined the 1960s, not as a time-restricted fashion, but as the beginning of an increase in alternative life-styles and counter-cultures spatially manifested as separated communities. Indeed, exploratory research of 1,250 alternative communities demonstrates that, even though hippy communes have largely disappeared since the 1970s, new communities of urban dropouts have begun to emerge based on ecological beliefs, religious ideologies or sexual preferences.

The Danish eco-village of Svanholm was established in 1978 and has since become a place shaped by ideals concerning ecology, income sharing, communal living and self-government, whilst rejecting mainstream society. Other examples are small, separated communities established by lesbian feminists in the 1970s and 1980s in rural areas of the United States. Athol in New York, Lavender Hill in California and Rootworks in Oregon are but a few of the 'Lesbian Lands'. These communities were an attempt to escape from a 'patriarchal society'. Rural areas were seen as particularly suitable places, because of notions of women's closeness to nature. In addition, spatial isolation in rural areas made it easier for the women to be self-sufficient, pure in their practices, and live in harmony with nature (Valentine, 2001).

Such communities of urban dropouts have, in the meantime, been classified as 'intentional', which simply means that they are characterized by a deliberate attempt to realize a common, alternative way of life outside mainstream urban society (Poldervaart, 2001). Members of intentional communities are united by a feeling of solidarity, which is often but not necessarily based on an all-inclusive common ideology, or common norms and values. For example, solidarity can be based on the rejection of a common enemy, as is demonstrated in the case study discussed in this chapter. People may join an intentional community because, for example, they reject the inequities of the existing system and long for a society committed to different ethics (Infield, 1955). In that case, joining can be an enriching and liberating experience for the community members.

When moving, new social identities develop as a new place (a new home) and social relations are explored and accommodated. Such a displacement remains liberating if these experiences are positive. However, several developments can reverse enriching experiences. For example, power relations that were previously contested may re-emerge, new members may challenge the established order of the community, the world outside of the community may cease to be a 'threat', or it may pose a new kind of threat. All of these developments can become a cause of dissatisfaction for community members (see below) and they can begin to feel out of place once again.

In the case of the artist community on the island of Usedom, the perceptions of political changes over time are of particular importance. The community members developed a unifying place identity, through identifying against first Nazi and later communist politics. Their identities were marked largely by resistance against the political ideology of the dominant part of society. In addition to creating a place for 'political sanctuary' on Usedom, the setting as a 'rural idyll' gave them space for artistic and intellectual activities that was simply impossible to claim in any other place in the East Germany of that time. However, times changed and Usedom did not remain unaffected either. The political transition to capitalism and increasing flows of tourists spoiling the 'rural idyll' renewed feelings of displacement with the community members. They could no longer identify with the community, which caused the community to disintegrate gradually.

Usedom: The Life History of a Place

Although the artist community no longer exists, it provides an interesting example through which to discuss the conflictual relationship between place and identity over time. The community in Mecklenburg Westpommern, Germany, on the island of Usedom (Figure 11.1) was studied in June 2002 as part of a wider study on intentional communities (the quotations cited below originate from that study). Although Usedom has been attracting artists sine the last decades of the nineteenth century, the area around the Waldstraße had become a 'hotspot' for artists from

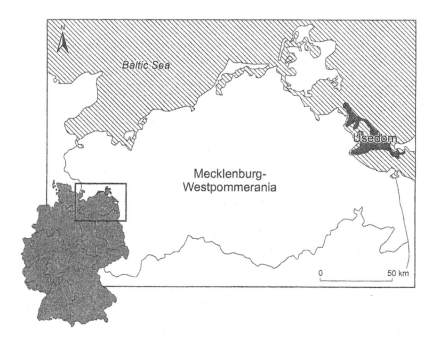

**Figure 11.1 Location of Usedom in Germany and Mecklenburg-
            Westpommern**

the early 1930s onwards. In the history of the community, different phases can be identified over time. These phases can be linked to Eastern Germany's political heritage of Nazism, communism and capitalism. The life course of the community as perceived by the artists reflects the relation between personal identities, social identities and place identities.

Foundation of the Community

The painters Otto Manigk (1902-1972) and Herbert Wegehaupt (1905-1959) settled in the Waldstraße in Ückeritz in 1932 (Lichtnau, 1993). They built their houses as summer residences, in contrast to their urban homes in Berlin. Karen Schacht was the third painter who settled in the Waldstraße, in 1940. In 1932, Otto Niemeyer-Holstein (1896-1984) also moved to the island of Usedom. Initially, he used the place as a summer residence in order to escape the city. Wegehaupt describes the island as a 'rural idyll', in which nature and all humans are united in harmony:

> The originality of the landscape, the beach and the sea continues to attract painters to the island in the summers. Here, the farmers cut the grain with the wide swings of their scythes, hay-wagons rambled over the sandy roads, the fishermen peacefully rowed onto the Baltic Sea. It was almost as in biblical times. A magically soft world, in which the painters mingled undisturbed as observers (translated from Wegehaupt, 1994, p. 4).

Although the painters initially sought temporary retreat on the small, relativelty isolated island, the Usedomer artists separated themselves from the influence of the state and started to live on the island throughout the entire year after Hitler reduced artistic freedom in Germany:

> ... the political leaders, that basically separated the country (Germany) from the world [...] and what should one do, when he is a poor painter? One retires, goes to the countryside, somewhere where he is least affected [by government policies] and does his thing.

The painters created an intentional community. Feelings of internal displacement that resulted in the physical displacement from the city (Berlin) to the countryside (Usedom) were both evoked by a sense of attraction to a 'rural idyll' and by a rejection of the fascist political regime. After the move, feelings of liberation prevailed. The identity of the community was based on the rejection of the political regime, embracing of a common political heritage and a shared sense of place.

Communist State Control

After the difficult early years of rebuilding following World War II, a feeling of freedom, belonging, solidarity and social engagement reunited the community. The artists resumed their work, and were mostly prepared to serve the social purposes of art in the early 1950s until the communist government initiated its so-called *Kulturpolitik* (Cultural Politics). Artists had to become a member of the *Verband Bildender Künstlern* (Society of Expressive Artists) and, only then, were granted paid assignments by the state. Although the artists were initially supportive of socialist ideologies, they were now expected to advance the building and glorification of the socialist state, which made them reluctant to take up assignments. As a result of this state pressure, they began, once more, to appreciate the spatial separation from Berlin, the centre of political power. The place identity of Usedom, already associated with the 'free' artists, again became important and the communal identity became based on identifying against a political system, albeit a different one (Van der Dollen, 1996). Their quiet resistance attracted respect and admiration by younger artists and, in the 1950s, Suzanne Kandt-Horn, Manfred Kandt, Rosa Kühn, Fritz Cremer, Rolf Werner and later Vera Kopetz, settled on Usedom.

It soon became obvious, however, that despite being united as a community in their identification against the communist political regime, the older artists gave meaning to their retreat in different ways. Schacht used it as a stepping-stone to West Germany, Kandt created a niche in his art by complying, and Niemeyer-Holstein sold many works privately from here using his connections with the rich elite of the GDR. The younger members of the community resisted the socialist regime more fiercely and attempted to free themselves from the restrictions that it imposed. Due to these different attitudes, some tensions arose in the community.

Inward/Outward Orientation

It is interesting to note that in addition to a differentiation between generations in terms of inward orientation, outward orientation was also different. Whilst the community lived separated from its national government, a strong but selective outward orientation dominated. In the 1930s, the artists were internationally oriented, both in art and in their other interests. They had their studios in Berlin, and had studied and travelled around Europe. Their art formed a distinctive contribution to European Modernism. The Nazi regime, however, made international contacts difficult, if not impossible as Germany became increasingly isolated from the rest of the world. Under communism, the artists established many contacts with intellectuals elsewhere in the GDR. They highly valued the – often political – conversations with them and their orientation towards the West, as the following quotation shows:

> Servais [a Belgian painter] lived in East Berlin, but he could travel to West Berlin, ... 'a man who could go through the Wall'. Consequently, he had a wider perspective. His visits to Ückeritz and working with him, were very interesting and exciting for me.

For some, the desire to be a full part of the international art community grew stronger and eventually estranged them from their own community. They suffered from limited opportunities in their own community, and started to identify against what they now regarded as a 'provincial' place. They idealized the parts of the world they could not visit:

> You would be happier under a bridge [in Paris] than over here in a house.

During communism, the restrictions on outward orientation were particularly frustrating for the younger painters, who had not known the freedom that the older generation had experienced and enjoyed before the coming to power of the Nazis. This observation highlights the importance of a sense of time in the development and relevance of this intentional community.

Differences Between Generations; Changes Over Time

As indicated above, from the late 1950s, difficulties and differences increasingly divided the community. A key factor in this was the distinction between two generations of artists in the community. The founding generation of the artists can be distinguished from the second generation, consisting of their children and students, who developed as artists during GDR times.

The difference in experience, combined with some youthful resistance or rebellion, may have resulted in a difference in attitude towards the state during the GDR regime. The first generation, most of whom died before the *Wende* (end of the communist regime) in 1990, showed more resignation and worked quietly and determinedly, even if adverse to the state norm. They passed on their critical, reflective attitude to the second generation, which that generation perhaps demonstrated in a less silent way:

> once, I was supposed to make a decorative mural with 'cheerful scenes from life in the State army'. My refusal to undertake the assignment was a motive for the Stasi to investigate the matter.

By the time the second generation started to oppose the socialist regime, the first had increasingly resigned. The decrease of their strong reflective and critical attitude intensified previously existing conflicts. As the two generations had experienced the community at different times, they equally perceived the place in different ways. Multiple identities developed within the community, based on different political heritages. However, in spite of differences in artistic styles and age, a sense of community still remained. Although the members of the younger generation all went away from the community for a while, for example to study in Berlin, most of them returned to Usedom. Oskar Manigk and Matthias Wegehaupt, artists and sons of Otto Manigk and Herbert Wegehaupt, still live in Ückeritz today.

German Reunification

The process of the ideological downfall of the GDR began relatively early in artistic circles. Already in the 1980s, many Usedomer artists provoked critical questions about environmental damage, disdain of cultural values and social structures through their paintings (Lichtnau, 1993). With this, the feeling of solidarity that was based on the projection of the socialist regime as a common 'enemy' disappeared:

> During socialism] they had the same problems, the same fears, the same needs ... and that bound them together. An artist had an oppositional function. That has become totally irrelevant nowadays. Back then, they really felt they could contradict, in particular while it was very much suppressed, they really have struggled.

After the *Wende*, the restrictions on travel were lifted and the second generation too, took the opportunity to travel to the West and integrate into the global art scene. Furthermore, the commercial aspect of art increased, which resulted in fierce competition and reduced the feeling of solidarity in the community. The rural idyll as a reason for settlement on Usedom that existed until the 1950s had already slowly disappeared, since Usedom was one of the most important tourist destinations in the GDR. Although the future of the community is uncertain, the interviews suggest that disintegration will continue. The first generation has largely died, whilst members of the second lead more secluded, individualized, lives. The artists who currently live on the island no longer speak of a community as such:

> But like back then, the artist community, that is gone ... they don't exchange ideas like before.

This observation appears to be a common aspect of post-communist transition as the unity in former communist states, created in the struggle against communism, has disintegrated and been replaced with increasing individualism and a multiplication of identities.

Conclusion

In discussing the experiences of place identity by an artist community in East Germany, we noted that 'identifying against' a common enemy was more important for the survival of an intentional community than a positive sense of place alone. This is not to say that place is not essential. Particularly in the formative phase, the association of positive features with a specific place, such as a rural idyll and a suitable, safe setting for a common way of life, are decisive in choosing where one wants to be, at what time and with whom. In cases such as ecovillages or 'Lesbian Lands' discussed above, the significance of place for the longevity of intentional communities may last. An ecologically oriented lifestyle, for example, appears pointless without specific environmental resources and spatial freedoms that can be more readily realized in rural areas. Equally, the closeness to nature and the distance to patriarchal contraints as sought by lesbian communities are likely to be place dependent. In contrast, the example of the artist community on Usedom illustrates that once the common 'enemy' (Nazism or communism) disappears, or becomes a 'friend' (for whatever reason), the reason to exist comes to an end. Compared with the eco- or lesbian communities, attachment to and relevance of place is considerably weaker. One may feel out of place or even restrained by the local context. Indeed, this proved to be the case in this study. As the local community disintegrated, the artists appeared to take advantage of new opportunities to integrate in other 'communities' of larger, even global, and multiple networks. These networks substitute the local community. Interestingly, involvement in such networks is largely detached from both place and time.

Acknowledgements

The authors would like to thank Tamara Kaspers-Westra for her cartography.

References

Cloke, P. and Little, J. (1997) (eds), *Contested Countryside Cultures: Otherness, Marginalisation and Rurality*, Routledge, London.

Cloke, P. , Milbourne, P. and Widdowfield, R. (2001), 'The Geographies of Homelessness in Rural England', *Regional Studies*, **35**(1), pp. 23-37.

Dear, M.J. and Wolch, J.R. (1987), *Landscapes of Despair: From Deinstitutionalization to Homelessness*, Polity Press, Cambridge.

Dollen, I. van der (1996), *Die Usedomer Maler. Landschaft 1933-1995*, KAT, Bad Honnef.

Duncan, C.M. and Lamborghini, N. (1994), 'Poverty and Social Context in Remote Rural Communities', *Rural Sociology*, **59**(3), pp. 437-61.

Hatty, S.E. (1996), 'The Violence of Displacement', *Violence Against Women*, **2**(4), pp. 412-28.

Hetherington, K. (1998), *Expressions of Identity: Space, Performance, Politics*, Sage London.

Hörschelmann, K. and B. van Hoven (2003), 'Experiencing Displacement - the Case of (Former) East Germany', *Antipode*, **35**(4), pp. 742-61.

Infield, H.F. (1955), *The American Intentional Communities*, Glen Gardner Community Press, Glen Gardner.

Lichtnau, B. (1993), *Usedom als Künstlerinsel*, Galerie Verlag, Fischerhude.

McHugh, K.E. (2000), 'Inside, Outside, Upside Down, Backward, Forward, Round and Round: A Case for Ethnographic Studies in Migration', *Progress in Human Geography*, **24** (1), pp. 71-89.

Mitchell, D. (2000), *Cultural Geography: A Critical Introduction*, Blackwell, Oxford.

Perry, R., Dean, K. and Brown, B. (1986), *Counterurbanisation: Case Studies of Urban to Rural Movement*, Geo Books, Norwich.

Poldervaart, S. (2001), 'The Concepts of Utopianism, Modernism and Postmodernism, Community and Sustainability', in S. Poldervaart, H. Jansen and B. Kesler (eds), *Contemporary Utopian Struggles: Communities Between Modernism and Postmodernism*, Aksant, Amsterdam, pp. 52-74.

Robertson, G., Mash, M., Tickner, L., Bird, J., Curtis, B. and Putnam, T. (eds) (1994), *Travellers' Tales: Narratives of Home and Displacement*, Routledge, London.

Russel, C. and Schofield, T. (1999), 'Social Isolation in Old Age: A Qualitative Exploration of Service Providers' Perceptions', *Ageing and Society*, **19**, pp. 69-91.

Rutherford, J. (ed.) (1990), *Identity. Community, Culture, Difference*, Lawrence and Wishart, London.

Samers, M. (1998), 'Immigration, 'Ethnic Minorities', and 'Social Exclusion' in the European Union: A Critical Perspective', *Geoforum*, **29**(2), pp. 123-44.

Teather, E. (ed) (1999), *Embodied Geographies*, Routledge, London.

Valentine, G. (2001), *Social Geographies: Space and Society*, Pearson, Harlow.

Wegehaupt, M. (1994), *Matthias Wegehaupt. Malerei, Grafik*, Galerie der modernen Kunst, Stettin.

Chapter 12

Literature and the Constitution of Place Identity: Three Examples from Belfast

Jonathan Stainer

Introduction

This chapter is an investigation into the competing senses of place expressed in three near-contemporary accounts of the city of Belfast during Northern Ireland's 'Troubles'. Using the examples of three literary 'fictionalizations' as source material, the focus is on place-determined constructs of sectarianism and nationalism. It argues throughout that bi-polar, sectarian constructions of place and identity, set within the discourse of rigidly territorialized rival nationalisms, have been (and still are) at the root of conflict in Northern Ireland. This pervasive sectarianism influences an unproductive, sterile, 'zero-sum' political culture, and is deep-rooted, powerful and institutionalized despite its arbitrary, irrelevant, anachronistic and illusory/imagined nature. Literary fiction is theorized here as a constituent of circuits of culture, in which the qualitative themes of place-cultures are both reflected and constructed. Literature both witnesses/reports place and interprets/re-constitutes it. The chapter analyses Eoin McNamee's *Resurrection Man* (1994), Mary Costello's *Titanic Town* (1992) and Ciaran Carson's *The Star Factory* (1997), in order to investigate sectarian identity politics, and to contest and suggest alternatives to this sterile antagonism.

The use of literary texts by geographers to uncover qualities of place identity is a reasonably well-established but not entirely unproblematic development. It stems from a desire to connect values, meaning, experience and the human subject to the understanding of places, partly as a reaction to the prior dominance of logical positivism within the discipline, and can also be positioned within a post-structuralist climate, which encourages interdisciplinary and discursive translation. The strengths of this approach lie in the insights revealed on human agency and senses of place, and the kind of contested geographical concepts which underlie belonging, politically and socially effective forces, and of course, conflict (Brosseau, 1994; 1995). Literature reveals inscribed identities, particularities and ideologies – it is not just spatial but 'placial'. The novel, Duffy (1997) argues, allows a more complex envisioning of the diverse and contested nature of place culture. For Daniels and Rycroft (1993), it presents different and competing experiential geographies, boundaries, perspectives and horizons.

Brosseau, however, is quite rigorous in identifying problems with the geographical-literary encounter. Literary texts are not *literalist* texts: they do not attempt or claim to represent the 'reality' of the world or even of human experience in the authoritative sense that geographical or social-scientific texts do. Geographers have too often used literary texts to answer previously determined research questions, ignoring the *formal* construction of the text, and the ways that various representative tropes narrate places (Brosseau, 1995). In other words, geography becomes the dominant discourse and literary theory is ignored (see also Sharp, 2000). Geographers therefore need to pay more attention to geographical and literary discourse itself, the contextual complexity of text as signifying practice. A desire for referential stability and a transparent conception of language may obscure what is truly different, radical or disruptive about literary texts. These can be active, resistant, and articulate places with a conditionality and plurivocality that contrast markedly with academic geography's monologism. By drawing attention to the production of meaning as an open-ended and interactive *process*, however, a more equitable dialogue between geography and literature might be usefully productive in criticizing or subverting enduring, essentialist narratives of the local and the national. It might allow new, 'positioned' ways of 'interpreting or representing social diversity and contingency in space', through the 'meeting of different voices and the confrontation of different logics' (Brosseau, 1995, p. 106 and p. 91). Sharp (1994, 2000) draws parallels between the modern novel and its association with the construction of the nation-state, a closed, bordered world where identities are stable and unchanging. Conversely, post-modern fiction formally challenges the genre and this particular view of the world, reflecting and producing instead the ambiguity and hybridity of the post-colonial world. Such work is 'multiply inscribed,' rejecting fixed and singular nationalisms. Further, geographers undertaking research with literature as the source medium need to acknowledge the background and, if possible, intentions of the author, as well as differences in reception, interpretation and response. In general, scrutiny is required of processes of reading and writing (see also Ryan, 2003).

Resurrection Man, *Titanic Town* and *The Star Factory*

One of a new generation of Irish novelists, Eoin McNamee's work displays something of a fixation with the psychopathology of sectarian hatreds and terrorist atrocities. *Resurrection Man* (1994) is a fictionalized re-telling of a real and particularly gruesome series of sectarian knife-killings perpetrated on Catholics by a Loyalist gang (the 'Shankill Butchers') in 1970s West Belfast. The narrative is relayed via the situated viewpoints of two principal protagonists, one a murderous paramilitary gang leader (Victor Kelly) and the other a Catholic journalist (Ryan). Despite its unrelenting bleakness and construction of a sinister and brutal narrative of place, I contend that the novel contains a veiled but powerful critique of sectarian forms of identification. It offers no *alternative* to sectarianism and grim 'Troubles' landscapes, but examines these subjects in detailed, critical ways. Crucially, it can be viewed as a sensationalist, exploitative text. Haslam (2000), for example, argues that its narrative seems complicit with the sadism of the killers, and does further violence to the real-life victims of the Shankill Butchers. It is frequently regarded as a postmodern and ambivalent work, divorced from any discernable ethical position. Johnson (1999) contends that in terms both of psychology and socio-political geography, the novel questions and destabilizes social-scientific explanations for violence. I argue here, however, that beneath the formal, detached cynicism, fatalism and postmodern shock-tactics, lies a critical examination of the nature of sectarian hatred and mistrust and a society divided by cultural-ethnic categorizations. The novel offers no overarching theories for the motivations of the killers, but hints obliquely at other possible explanations, accumulated knowledges and experiences that offer a particular insight into essentialism.

Memoirist Mary Costello was born in West Belfast in 1955, into a working-class Catholic family, her adolescence and teenage years coinciding with the onset of the 'Troubles' in the late 1960s and the serious unrest of the early 1970s. *Titanic Town* (1992) is a semi-autobiographical account of this childhood and adolescence. Costello substitutes her thoughts through the character of Annie McPhelimy, the novel's first-person narrator. The text is partly remembered and partly invented – for the author this is an appropriate parallel to the way in which, historically, 'truth and untruths, misunderstandings, myths and rumours...in Belfast, ferment into history,' a 'dismembrance of things past' (*Titanic Town*, Author's Note). The subject matter is the life of the Catholic-Nationalist community in the Falls and Andersonstown, working class districts of West Belfast. Costello juxtaposes the 'everyday,' common experiences of growing up with the 'extraordinary' events of the time, creating an insider's view of an IRA-controlled Catholic enclave. It details the difficulties associated with attempts towards non-violent dissent and political amelioration in the face of opposition from a community, which seems to some extent to be affiliated to paramilitary groups. As well as confronting political issues directly, Costello describes the limitations imposed on ordinary life for her teenage protagonist. The anecdotal writing style employed by Mary Costello in *Titanic Town* stands in some contrast to the morbid and oppressive prose of Eoin McNamee. Events, sometimes violent and disturbing, are relayed in a manner, which is comparatively ironic and often blackly humorous. If, formally, McNamee's writing can be seen as an extreme, severe and *dissatisfied* critique of politics and society in the North of Ireland, then Costello's work might appear, in contrast, lacking in such criticism. Far from being a symptom of complacency, however, this more subtle approach should perhaps be read as reminiscence and optimism in the face of adversity and seemingly insoluble conflict. The text contains constructive and insightful examinations of sectarian place politics, paramilitary violence and their constraining effects on society. It appropriately reflects this chapter's concern to identify the city's particular constructs and identities of place, its imagined geographies.

Ciaran Carson was born into an Irish-speaking family in Belfast in 1948. *The Star Factory* (1997) is a sequence of interconnecting essays, which picture and explore the city of Belfast, (de)constructing and revealing the form and qualities of its remembered and imagined spaces. His use of language is unconventional; as well as displaying 'a witty and playful regard for language', he is not afraid to make use of techniques of free association, digression and even contradiction (Welch, ed., 1996, p. 84). Further investigation of these techniques reveals a deliberate attempt by Carson to draw the reader's attention to the subjectivity and contingency of meaning in the textual representation of an external 'reality'. By subverting conventional literary forms, he underlines the ambiguity of language, and thus the multiplicity of different meanings not only within the text but also in the 'real world' beyond. This has led some critics to situate his later work in the context of postmodern critical theory and practice (Welch, ed., 1996). His prose conveys the subjectivity, memory and experience of places. Its modes of expression are fragmented and ephemeral, often focusing on minutiae and the seemingly obscure, and embarking on tangential distractions and digressions (after Donahue, 1997). This textual strategy, I contend, creates a sense of constant re-ordering and redefinition, and a resistance to closure, circumscription and stable categories. It subverts the fixed dualistic sectarian hegemony of the North.

Partisan Localism: Identity and the Cultural Politics of Place

It is argued here that place-based identity and senses of belonging in Northern Ireland, and elsewhere, are socially effective, enduring and problematic, but deliberately constituted, mediated, and in some sense, then, ultimately 'false'. Such exclusive and excluding narratives are perpetuated through *geographical* ideas, where places are constructed and reproduced (see, for example, Agnew, 1998; Duncan and Ley, 1993). Partisan localisms such as loyalism and republicanism are set within a broader idea of the nation, where the delineation of territory is conflated and (con)fused with idealized landscapes and culture (Anderson, 1988), and legitimized by particular (invented) linear historical narratives (Hobsbawm, 1991). British and Irish identities in Northern Ireland are defined for the most part in opposition to each other – a consequence perhaps, but also a cause, of conflict. Hegemonic Irishness has traditionally been constituted in terms of a Gaelic, Catholic, socially conservative, mythologized rural West, 'untainted' by modernity and industrialization (Johnson, 1993); this can be regarded as a historically situated (nineteenth century) and therefore anachronistic construction, designed to differentiate and distance the emerging nation from imperial and industrial Britain (see, for example, Loughlin, 1998). In turn, Ulster/ Protestant identity, other than being 'not-Irish,' has been a much more poorly articulated, less coherent and charismatic defining trope. Of course, the conflict itself has become characteristic of an idea of 'Northern Ireland,' but aside from this, political unionism has failed to produce a state-legitimizing narrative of place, relying instead on incoherent notions of Britishness and often negative or archaic ethno-cultural markers of identity such as Orangeism (Graham, 1994, 1997). Political interaction and organization in Northern Ireland is almost entirely reduced to and dominated by ethno-nationalist identity, British/Protestant and Irish/Catholic (although there are further fissures and fragmentations within these, not least due to class) zero-sum sectarian tribalism and the border issue. Again, this can be seen as a consequence of 35 years of conflict, but emotive, effective and yet invented, distorting, mediated and imaginary constructs of identity and place only serve to *perpetuate* conflict, and must be challenged if a sustainable political resolution is to be achieved.

These dominant representations appear as facts, 'rendered common-sense,' and 'discursively naturalised' (Rose, 1994, pp. 46-7). However, it can be argued perhaps, that such sets of meanings are arbitrary and contingent, rather than secure, unchanging and inevitable (Barnes and Gregory, 1997). Places should be articulated in more fluid, perspectival and inclusive ways, 'progressive and dialogic' (Nash, 2000, p. 28), 'less-bounded, more open-ended' geographies (Driver and Samuel, 1995, p. vi). Massey (1994) stresses that place is always open and provisional, while for Soja (1996; 2000) it is hybrid, consisting of the real and measurable as well as the imagined or socially constructed. Bhabha (1994) also makes a case for the hybridity of place, asserting that culture and identity in the postmodern, postcolonial world are formed in 'in-between' spaces, neither one thing nor another. He envisages new 'anti-nationalist histories' through which 'the politics of polarity' might be avoided (1994, pp. 38-9). Crucially, however, identities tend to resist hybridization; 'cultural racism' is a 'pervasive force' (Werbner, 1997, p. 3). For Sharp, the uncertainties and threats (real-and-imagined) of transcultural environments drive groups and individuals to seek self-definition through the non-negotiable absolutes of nation and religious/cultural traditions (1994, see p. 74). Such essentialisms might be resisted through awareness of the discursive and semiological systems (as well as performances and processes) that create places, and through ambivalent geographies and identities which do not (formally) perpetuate established constructions (Sharp, 2002). According to Hooper (2001, p. 704), the 'most important task [is] to investigate how the (real-false) borders we live with and incorporate – those particularizing differences and divisions... – are produced.'

Narratives of Place in the Three Fictionalized Texts

The remainder of this chapter consists of a concise interpretative reading of the texts with regard to sectarian and national themes (as well as their formal construction). The novels inform an analysis which is critical and disruptive of some of the more 'taken-for-granted' sectarian narratives, questioning their legitimacy or validity. The analytical section is divided into four themes. In the first, sub-nationalist identity constructs are 'de-centred' and the dissonance between the circumstances of their creation and present context evaluated. Secondly, it is argued that sectarianism uses *signs* which are contingent despite appearing 'natural'. The third theme explores the vicious circle of 'real-and-imagined' sectarian violence, spatial segregation and mutual ignorance, the interplay of measurable phenomena and those of the socio-cultural imagination. Finally, narratives of perspective-enriched hybridity and reconciliation are outlined, and a case is made for the contingent and 'multiply-situated' nature of identity and the city. The former three themes draw on *Resurrection Man* and *Titanic Town*; the latter on *The Star Factory*. It is argued throughout that the existing sub-nationalist affiliations are an anachronistic and problematic hindrance to a peaceful, integrated society, and that more open-ended, less rigidly territorial alternatives are necessary for the grounding of enduring political structures.

De-centring

Enduring, rival nationalist narratives, then, inform Northern Ireland's sectarian geographies. These narratives, however, are 'de-centred,' and divorced from their original context and agency to become 'natural,' fixed 'truths'. The sign becomes the thing itself, the terms of a narrowly construed debate. This uncritical observation of 'timeless' cultural-political convictions can be discerned in both the ultra-modernist/rationalist absolutism of McNamee's loyalists and Costello's republicans. Uncompromising nationalism encourages sectarian thoughts and actions despite their abstraction and irrelevance.

　　　The Belfast depicted in *Resurrection Man* is defined by elements such as work-ethic, purpose, progress, modernism and industrial materialism, and can thus perhaps be positioned within a representative trope of Britishness which stands in opposition to idealized rural constructions of Irishness. The city is characterized by its nineteenth-century origins in the era of Empire and fundamental certainties. McNamee ascribes to it a 'sombre mercantile ethic' (p. 85), and 'sectarian and geographic certainties' (p. 17). Such ideas are exaggerated to a form of fundamentalist super-rationalism in the minds of *Resurrection Man*'s protagonists, a murderous and psychotic anti-Catholicism. For gang leader Victor Kelly, the killing of the Catholic Irish other is a moral duty and gives a sense of purpose. The constructs are naturalized, unchallenged and seemingly beyond argument or critical examination (Cosgrove and Daniels, 1988; Duncan, 1993). This trope of identification is characterized by modernity and certainty, and is itself a modern (and moral) certitude.

Titanic Town vocalizes other positions within a West Belfast often portrayed as a conformist republican monolith. Many of these voices, and the overall tenor of Costello's text, are dissenting, critical of this politically dominant form. This novel, I argue, is a satirical intervention, an undermining of the simplified yet enduring ideas propagated by militant activists. The 'timeless' cultural nationalism which underlies popular republicanism is portrayed as sentimental and uncritical. Costello (p. 43) facetiously attacks the 'cultural politics of place' (Rose, 1994, p. 46), where politics, culture and nationalism are seamlessly fused and confused, where the '...country's revolutionaries are also its poets.' She also seems to be questioning and ridiculing the unrigorous and emotive nationalism which surrounds the protagonist.

The conflict seems to be informed by modernist nationalisms, with Belfast as its urban focus. There is perhaps a tension between the identity of the city as industrial, modern, Victorian and so on (British), and of idealized versions of Irishness in a range of guises from rural 'pre-modernity' to postmodern commodity. The city depicted in *Resurrection Man* is a functionalist, purposeful place and, although prosperity has long since departed – 'there is a sense of collapsed trade and accumulate decline,' (p. 3), industry is still a defining trope. It could be argued that place identity is to some extent then characterized by industrial 'emasculation,' combined with (supposedly) 'modern' and 'British' attributes such as progress and certainty of purpose. This dissonant combination of chronic insecurity (in economic terms) and enduring, effective, but *de-centred* identity constructs may result in a kind of collective psychosis – locally and individually, due to a loss of function and livelihood, of (male) power and the international standing of the nation. This may be over-compensated for by forms of extreme localism and nationalism.

Deconstruction

Here, it is argued that social-sectarian categorization, such as the profound anti-Catholic hatred displayed by *Resurrection Man's* protagonists, whilst effective, is merely symbolic and illusory. This construction seems to be qualitatively informed at least in part by fundamentalist Protestant religion, 'merciless theologies' (p. 37) where 'Catholics were plotters, heretics, casual betrayers' (p. 9). For McNamee, this sectarian othering is not confined to extremists, it is all-pervading and institutionalized, an ambivalence that permeates all social classes, the police ('detectives hinted at threads of sympathy for the killers in the lower ranks,' p. 146) and politics ('there seemed to be a dark current of approval in the political sphere,' p. 146). Sectarianism is also 'discursively naturalised' (Rose, 1994, p. 46), so that it justifies violence and intimidation, as in the following example. The protagonist's father, James Kelly, suffers intimidation in his Protestant workplace (the shipyard) on the basis of his 'ambiguous' surname – 'Kelly' is not an exclusively Catholic or Protestant marker and therefore an unreliable token of identity. I argue that McNamee uses this uncertain signifier to underline the tautological and arbitrary yet socially effective nature of categorization. Kelly is singled out on the grounds of his Catholicism, which is uncertain, and (crucially) never stated in the text – a deliberate ambiguity. He survives by 'positioning himself against the unstable surfaces offered by the world' (p. 167). What is important to his persecutors is the *categorization of the individual* rather than the individual. The discourse of the sign becomes the thing itself. The text effectively deconstructs the 'myth of realism,' the illusory narratives of fundamentalism, seeming to draw attention to their 'unstable surfaces' and the contingency and social construction of meaning. Kelly is assigned a category on the basis of a (neutral, ambiguous) name/symbol, and social meaning is seen to be deliberately assigned rather than inherent. In this, and in many other examples within the text, McNamee presents sectarian categorization as a discursively naturalized system of representative symbols.

Real-and-Imagined Sectarian Division

The spatial division of communities along sectarian lines, despite its basis in culturally constructed forms, is a real and measurable phenomenon. Sectarian affiliation legitimizes violence and division and thus facilitates its own reproduction. Institutionalized and politicized sectarianism allows the two 'communities' to become mutually isolated and alienated. Conflict results, perhaps systemically, from this mutual ignorance, and reductive mediations and representations of the other. For McNamee, a failure of understanding between protagonists becomes a metaphor for the mutual misunderstanding of differently categorized groups. It is suggested here that the breakdown of communication between members of the Kelly family – 'sooner or later…their conversations started to diverge' (p. 72), 'like a silent allegation of blame or deceit' (p. 73) – is a representation of the gulf of incomprehension between isolated and solipsistic political-cultural imaginaries. Such forms of identification only make sense within the terms of their own particular discursive and representational systems; they fail to acknowledge or comprehend *each other*. Such dissonant place narratives are each mutually unintelligible from sited positions of interpretation within the other. This sets up a cyclical process where segregation reinforces the reproduction of incompatible cultural ideas, which in turn sustains separateness, suspicion and lack of interaction. The impossibility of communication between the protagonists in *Resurrection Man* appears as a comment on the mutual solipsisms of sectarian antagonism.

Both McNamee and Costello characterize Belfast as being physically divided into discrete territories, effectively unknown and foreign to each other. When Victor Kelly drives his girlfriend Heather along the internal borders that separate Protestant and Catholic enclaves, she senses that the latter places are forbidden, part of another, almost mythical place, despite being part of her home city. For example: 'she had never been this close before although she had seen these places on television' (p. 45). Costello's city is crossed by 'peace lines' which divide exclusively Protestant and Catholic areas in a blunt reminder of the real separation engendered by cultural constructs. Territory is bounded, divided and (paramilitary) power wielded accordingly therein, recalling nation-states in microcosm. Social and political identities are highly territorialized, which is, in a sense, representative of the Troubles and of national division. This spatial delineation, of course, is more than mere abstraction – it is overwritten with violence, the threat of violence, and real social effects. Bernie, the mother of *Titanic Town*'s protagonist, has a job in a linen mill in a Protestant enclave (at the height of paramilitary campaigns of violence) and thus must cross the sectarian interface daily as an economic necessity. The workplace is rife with sectarian tension – 'remarks would be passed by the Protestant majority, allegations would be whispered' (p. 86) – and intimidation eventually forces the Catholic workforce to resign: 'it became obvious, even to mother, that it would end badly' (p. 89). So here, division is 'real-and-imagined' (Soja, 1996; 2000) or 'real-false' (Hooper, 2001), with real, polarizing, isolating social and economic consequences (due to arbitrary and imagined constructions of identity).

Hybridity

Whilst Carson allows also that violence and division are defining characteristics of the city of Belfast, he appears to go further, even to the point of seeking to transcend that division. He uses various narrative devices to do this, such as imagining the bridges between the east and west of the city as a kind of metaphor for reconciliation and renewal (p. 237), or characterizing the 'fragile souvenirs' of traditional cultural-national symbolism as transient and impermanent (p. 257). More broadly, he sees the city from new angles and in different configurations, alluding frequently to hidden, underground spaces – for example, 'a largely inaccessible...narrative dimension' (p. 117, also pp. 151 and 210) – marginal or parallel geographies which seem to resist or refuse to recognize the dominant constructs of zero-sum politics.

Carson articulates personal and place identity in terms of two national cultural entities normally presented as exclusive and dissonant. An 'official,' almost colonial Britishness is represented by the medium of radio, and by the BBC in particular. This Britishness is embodied in the establishment and the state, and contrasts with Carson's Catholic, Irish-speaking domestic upbringing. Both 'conceived nations,' however, are encompassed without antagonism in his experiential narrative, where a mix of influences is recognized. He feels 'complicit' with the decline of the paternal British authority represented by the BBC World Service, and yet comforted by its presence (p. 104), suggesting acceptance of the spatial coinciding of dissonant political-national ideas. This is, in a sense, a hybrid space, a rapprochement between the 'radically different senses of place' (Driver and Samuel, 1995, p. vi) of rival nationalist narratives of belonging.

As well as the subterranean concept mentioned above, Carson also uses one of elevated perspective to stress the positionality, subjectivity and plurivocality of the urban experience, and to create a sense of distancing which alludes to new kinds of spaces. An array of different viewpoints allows a kind of 'double-situation,' where more complex geographies, competing fields of experience, might be represented. From elevated viewpoints around the city, usually the mountains to the west, he inserts distance between the individual and their imagined 'self-in-place,' allowing the subject to view itself from afar, simultaneously inside and outside, observer and observed. The mountains are also more generally expressive of freedom, of escape from the oppressive social norms of the sectarian city (see, for example, 101). The most notable example of this elevated trope is in the 'bi-location' of the 'Invisible Boy' (p. 203), a practical joke remembered from Carson's schooldays, in which one boy exits a second-storey classroom window and climbs precariously along a ledge, staying out of view of the teacher for the duration of a lesson. For those remaining in the classroom, the Invisible Boy becomes 'an *alter ego*, and through the Boy each boy could experience *bi-location*' (my emphases, p. 203). Carson recounts: 'I was once a Boy myself...I recalled the dizzy sudden view I got as I first gazed through the open window...I fell into a reverie of West Belfast...I felt the verges of a flying dream' (p. 204). As a metaphor, it suggests that one might regard/understand the self more 'objectively' through self-reflexive distancing, stepping 'outside', and creating new perspectives. One might thus appreciate the plurivocality of the city and resist its sectarian orthodoxies. The 'Invisible Boy' is simultaneously inside and outside, excluded and included, mysterious, obscured, hidden, and yet familiar and universal. Carson produces a 'dialectic spatiality' with which to contest the partisan localisms of sectarian place.

Conclusion

In this chapter, I have argued that excluding, territorialized senses of place, drawing on nationalism and localism, underlie politics and conflict in Belfast. The critical engagements with the literary texts inform the discussion in several useful ways. The texts stand outside the closed and enclosing symbolic system of sectarian place politics and serve to critique and de-centre their content. Place, culture and political praxis are seen to be conflated as a fundamental, imperative certainty, based on emotive, effective though anachronistic representations. Sectarianism is an assumed 'fact,' despite its temporal and social-spatial contingencies and inconsistencies. Its social meanings, deliberately assigned, might also be deliberately destabilized. Crucially, however, the 'reality' of segregation facilitates a cycle of ignorance where sectarianism (and attendant violence) reproduces itself. The constructs are 'real-and-imagined'. Sectarian institutions, narratives and geographical patterns are socially, politically and economically limiting and damaging in measurable ways. The perspective, contingency and positional duality of the final reading suggests a need or desire for a more inclusive and 'progressive' conceptualization of social space, descriptively and theoretically, one which perhaps questions boundaries, categories and discursively naturalized arbiters of difference.

References

Agnew, J. (1998), 'European Landscape and Identity', in B. Graham (ed.), *Modern Europe: Place, Culture, Identity,* Arnold, London, pp. 213-35.

Anderson, J. (1988), 'Nationalist Ideology and Territory', in R.J.Johnston, D.B. Knight and E. Kofman (eds), *Nationalism, Self-determination and Political Geography*, Croom Helm, London, pp. 18-39.

Barnes, T. and Gregory, D. (1997), *Reading Human Geography: The Poetics and Politics of Inquiry*, Arnold, London.

Bhabha, H. (1994), *The Location of Culture,* Routledge, London.

Brosseau, M. (1994), 'Geography's Literature', *Progress in Human Geography,* **18**, pp. 333-53.

Brosseau, M. (1995), 'The City in Textual Form: *Manhattan Transfer*'s New York', *Ecumene*, **2**, pp. 89-114.

Carson, C. (1997), *The Star Factory*, Granta, London.

Costello, M. (1992), *Titanic Town: Memoirs of a Belfast Girlhood*, Methuen, London.

Daniels, S. and Rycroft, S. (1993), 'Mapping the Modern City: Alan Sillitoe's Nottingham Novels', *Transactions of the Institute of British Geographers,* **18**, pp. 460-80.

Donahue, M. (1997), 'Biography - Ciaran Carson', in Gonzalez, A. G. (ed.), *Modern Irish Writers: A Bio-critical Sourcebook*, Aldwych Press, London, pp. 40-3.

Driver, F. and Samuel, R. (1995), 'Rethinking the Idea of Place', *History Workshop Journal,* **39**, pp. vi-vii.

Duffy, P. (1997), 'Writing Ireland: Literature and Art in the Representation of Irish Place', in B. Graham (ed.), *In Search of Ireland: A Cultural Geography*, Routledge, London, pp. 64-83.

Duncan, J. and Ley, D. (1993), 'Introduction: Representing the Place of Culture', in J. Duncan and D. Ley (eds), *Place/CultureRrepresentation*, Routledge, London, pp. 1-21.

Graham, B. (1994), 'No Place of the Mind: Contested Protestant Representations of Ulster', *Ecumene,* **1**, pp. 257-81.

Graham, B. (1997), 'Ireland and Irishness', in B. Graham (ed.), *In Search of Ireland: A Cultural Geography*, Routledge, London, pp. 1-15.

Haslam, R. (2000), "'The Pose Arranged and Lingered Over': Visualising the 'Troubles'", in L. Harte and M. Parker (eds.), *Contemporary Irish Fiction: Themes, Tropes, Theories*, Macmillan, Basingstoke.

Hobsbawm, E. J. (1991), *Nations and Nationalism Since 1780: Programme, Myth, Reality*, Cambridge University Press, Cambridge.

Hogan, R. (ed.) (1996), *Dictionary of Irish Literature*, Aldwych Press, London.

Hooper, B. (2001), 'Desiring Presence, Romancing the Real', *Annals of the Association of American Geographers*, **91**, pp. 703-15.

Johnson, N. (1993), 'Building a Nation: An Examination of the Irish Gaeltacht Commission

Report of 1926', *Journal of Historical Geography,* **19**, pp. 157-68.
Johnson, N. (1999), 'The Cartographies of Violence: Belfast's *Resurrection Man',* *Environment and Planning D: Society and Space,* **17**, pp. 723-36.
Loughlin, J. (1998), *The Ulster Question Since 1945*, Macmillan, Basingstoke.
Massey, D. (1994), *Space, Place and Gender,* Polity, Cambridge.
McNamee, E. (1994), *Resurrection Man*, Picador, London.
Nash, C. (2000), 'Historical Geographies of Modernity', in B. Graham and C. Nash (eds), *Modern Historical Geographies,* Longman, Harlow, pp. 13-40.
Rose, G. (1994), 'The Cultural Politics of Place: Local Representation and Oppositional Discourse in Two Films', *Transactions of the Institute of British Geographers,* **19**, pp. 65-81.
Ryan, J. R. (2003), 'History and Philosophy of Geography: Bringing Geography to Boot, 2000-2001 (Progress Report)', *Progress in Human Geography*, **27**, pp. 195-202.
Sharp, J. P. (1994), 'A Topology of 'Post' Nationality: (Re)mapping Identity in *The Satanic Verses'*, *Ecumene*, **1**, pp. 65-76.
Sharp, J. P. (2000), 'Towards a Critical Analysis of Fictive Geographies', *Area,* **32**, pp. 327-34.
Sharp, J. P. (2002), 'Writing Travel/travelling Writing: Roland Barthes Detours the Orient', *Environment and Planning D: Society and Space*, **20**, pp. 155-66.
Soja, E. W. (1996), *Thirdspace: Journeys to Los Angeles and Other Real-and-imagined Places*, Blackwell, Oxford.
Soja, E. W. (2000), *Postmetropolis: Critical Studies of Cities and Regions,* Blackwell, Oxford.
Welch, R. (ed.), (1996), *The Oxford Companion to Irish Liiterature*, Clarendon Press, Oxford.
Werbener, P. and T. Modood (eds), *Debating Cultural Hybridity: Multicultural Identities and the Politics of Anti-racism*, Zed Books, London, pp. 1-26.

Chapter 13

Imagining Newfoundlands

G.J. Ashworth

Introduction: Who Invented the Rock and its People?

This chapter attempts to apply some of the more general questions raised in this book to a specific place and time, namely the North Atlantic island of Newfoundland since its officially recorded discovery by Europeans in 1497. As many have noted, senses of place are constructed from various elements combined into place identities at various times for various purposes. This chapter attempts to describe these elements in their varying constructions and, in particular, the interactions between different imagined Newfoundlands. The assertion that place identities are ascribed prompts the immediate questions, 'who did this, when and why?' This matters in particular because a place image created and promoted for one purpose may interact with quite different images intended for different purposes and different markets. Specifically place images deliberately created for external consumption may become internalised, while external markets may adopt those intended for local consumption.

Newfoundland (Figure 13.1: spatially defined here as the island excluding its mainland 'colony' in Labrador, which has developed quite different identities) has three dominant and influential characteristics that render it a suitable case study. First, it is physically, ethnically and temporally bounded. The 'Rock', as its popular nickname suggests, has distinct physical characteristics of enclosed insularity and separation. Secondly, its people are almost exclusively of British origin: other previously present ethnic groups such as the Dorset Inuit, the Beothuk and Micmac Amerindians or the NorseVikings have played no formative role in the place identity. Thirdly, there is a clearly defined and, by North American standards, relatively long time span beginning with 'discovery' (1497) and annexation as the first colony of the British Empire (1583) (Matthews, 1973). Physical distinctiveness, cultural homogeneity and a lengthy history all make the study of the formation, evolution and interaction of senses of place more apparent.

On Fish, Fishing, Fishermen and Fishing Ports

Cod may have been 'the fish that changed the world' (Kurlansky, 1997) but in the process it created Newfoundland. The motive, pace, distribution and form of settlement in Newfoundland has been determined by the requirements of fish catching, processing and marketing. A particular settlement adaptation to a physical environment and to an economic function was required by the exploitation of inshore fish resources using hand-lines from dories ('jigging'), and more recently, of necessity, supplemented by inshore scallop and lobster fishing. This necessitated numerous small settlements along the indented coastline with water access to the fishing grounds and to the merchants and markets but lacking land access. Indeed, in terms of settlement history it would be more accurate to envisage Newfoundland as an archipelago of settlements, most of which was only created in the past 50 years, accessible over water rather than a single island with settlements linked by land access. The result was the *outport*, a settlement with a distinctive distribution pattern and physical form. This fish-dependent economy and settlement structure

Figure 13.1 The island of Newfoundland

produced a resourceful, self-reliant, interdependent society of small cohesive communities whose relative social isolation preserved and nurtured distinctive cultural characteristics, notably in linguistic usage, social behaviour and custom, artefact design and artistic expression, especially in music.

The outport has become the receptacle for the idea of Newfoundland. It is both the place where the 'real' Newfoundland is to be found but also is itself an artefact which has become the ultimate expression of 'Newfoundlandness'. This association between a particular settlement form and place image is now so powerful that, 'the 'real' Newfoundland is the outports and their people' (Overton, 1980, p. 106). So potent is the idea that, 'Newfoundland culture is out there in the outports and ... it has certain essential characteristics once and for all' (Overton, 1996, p. 17) that the outport has passed from mundane economic response to symbol of 'authentic' Newfoundland. History has become heritage which in turn has become identity. The irony is that, for whatever reason, cod-dependent Newfoundland has ceased to exist as an economic reality at just that moment in time when it has been

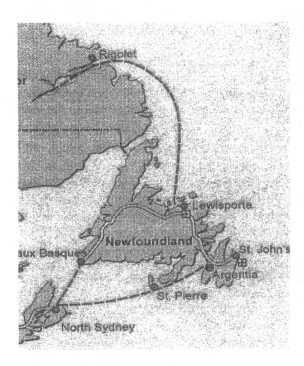

Figure 13.2 Newfoundland: accessibility

elevated to the central element of Newfoundland identity for consumption on and off the island. This in itself creates a number of dilemmas in the production, management and marketing of a Newfoundland dependent upon the outport of the imagination (Figure 13.2).

Some Imagined Newfoundlands

There are at least six identifiable elements in the place image of Newfoundland.

The Treasure Chest

The idea of Newfoundland as a cornucopia of resources, an inexhaustible treasure chest awaiting plundering, is perhaps the oldest externally projected image. The report of Cabot's famous basket that needed only lowering to be miraculously filled with cod is surely one of the oldest pieces of hyperbolic place over-selling, on a par with Leif Ericson calling his ice-bound discovery 'Greenland' and the L'Anse aux Meadows settlement, 'Vinland'. Place boosterism to attract settlers and investors has been extended at various times from fish and marine mammals to game, timber, stone, slate, and more recently gas and oil. The barren wilderness as void is transformed into a bounteous natural endowment. The 'modernization' programmes of the post-war Smallwood administrations strongly advanced this image of limitless resources waiting to be unlocked. The bonanza economy with prosperity for all 'just around the corner' resurfaces from the first European settlement to the present, perhaps as a compensatory counterpoint to the 'hard times' image discussed below.

The Norway of North America

Newfoundlanders throughout most of the history of settlement have seen wild nature as either contributing inconvenient experiences or obstacles to be traversed, endured or, if at all possible, removed, or as a source of directly consumable foodstuffs, building materials and tradable products. The construction of the railway and the opening of the scheduled ferry routes to Cape Breton in the 1890s allowed the commodification of the physical geography into the sellable product, 'wilderness'. External demand for this had been generated in Europe and the United States by a nostalgic romantic reaction to nineteenth-century industrialization and urbanization. Wilderness was to provide an aesthetically enriching, physically healthy and spiritually uplifting experience. The 'Norway of the New World' was constructed from the negative attribute of the absence of obvious human occupation and the positive contributions of scenery and wildlife. Wilderness is, of course, a cultural construction using space, scenery and wildlife and identification with wilderness is seen as providing cultural and spiritual advantages. However it has negative associations and certain practical disadvantages. The attractive feature 'wild' is easily translated as uncontrollably bothersome or, at worst, dangerous; 'adventure' and 'challenge' becomes insecurity and uncertainty; a pleasant 'solitude' becomes a distinctly inconvenient isolation. Wilderness has elements which may either support or undermine development strategies.

The Last Refuge of Pre-industrial Society

The parallel construction of wilderness from the physical geography was that of the simple fisherfolk from the human geography. A commonplace set of economic activities was self-consciously reinvented as the 'Newfoundland tradition' of simple 'fisherfolk'. This was a quite deliberate reinvention of Newfoundland, at least initially by outsiders, as the last stronghold of non-industrial values (Mowat, 1972; 1989). As with wilderness, this idea of a pre-industrial unspoilt civilization was a product of industrial urban societies in Europe and the United States. The 'noble savage' became the simple but wholesome fisherman. The economic occupation conferred not merely a livelihood but a way of life and most significant, a set of behavioural, moral and ethical qualities. The fisherman and his family are diligent, adaptable, reliable, honest, hospitable, courteous and non-violent. The association with the outport is not only strong, it is a *sine qua non*. In reality of course this idea was not only created by industrial society but it could only come into existence through an external validation and designation that in itself transformed the phenomenon into a self-conscious performance.

The Authentic Folk

Stemming in part from the above is the idea of folk culture, the fisherfolk experiencing wilderness and hard times expressing themselves in a timeless vernacular culture. The image is of small physically isolated, tight-knit communities, of necessity living and working together. They are in touch with the realities of nature, express their identities of survival and resistance through various vernacular traditions which the outport created and then preserved in a changing world. Folk is composed of a range of distinctive ethnological survivals, encompassing linguistic expressions, customs and artistic production. The last includes music, architecture, boat design and, more recently, design pottery, wood carving, textile and clothing and even gastronomy.

The dating and reasons for the creation of a Newfoundland folk culture have been debated elsewhere (see for example, Overton, 1988; 1996). In fact much of this is a quite recent development, although no less vernacular for that. Folk provides a clear marketable image and a highly flexible product range which was not dependent upon any physical resource base. This could be projected to both residents as legitimating their place identities and to visitors as a defined brand image, an on-site experience and a highly portable souvenir. Its success is demonstrable in the fixing of a clear and dominant global and local place image of old traditional folksy Newfoundland. The island has therefore locked itself in, over the long term, and would now find it next to impossible to escape into alternatives.

The folk product depends for its quality upon a perceived authenticity in three senses: authentic folk design and materials ('vernacularism'); authentic production process ('craftsmanship'); authentic producer ('craftsman'). Each of these is almost the antithesis of efficient, profitable, capitalist, mass production: the word 'almost' betrays the solution and its dangers, namely that the perception of such authenticity must be maintained despite its absence of reality. The intrinsic deception of what Cohen (1979) famously described as 'staged authenticity' involves a voluntary suspension of disbelief on the part of the consumer and an extreme skill and sensitivity on the part of the producers and promoters. It is not, as Stuckless (1986) seems to suggest for Newfoundland, that images should be truthful, which raises far too many questions about the nature of truth, but that the 'authenticity of the experience' has to be preserved.

Hard, Hard Times

'500 years of stubborn survival in the face of tyrannical oppression, inhuman laws, a harsh environment, invasion and pillage, holocausts, economic depressions and violent politics' (City of St. John's Tourist Commission, undated tourism map). One might wonder when confronted with this piece of hyperbole, not merely which tyrants, oppression, holocausts, invasions, violence and the like have afflicted Newfoundland, and passed unremarked by the rest of the world or by the attentions of historians, but how this supposed experience produced a host society characterized, on the same page, by 'affection, respect, humour and gentility'. More pragmatically, one might wonder why the tourism agency imagines that such a catalogue of woes is attractive to fun-seeking tourists who are thus invited to view the misery of the suffering victim. Clearly the adoption of the role of victim is for local consumption. Wilderness and folk combine easily in the idea of man in nature which becomes nature against man in a struggle for survival on 'the Rock'. To the unrelenting physical forces is added economic and political oppression.

If Newfoundlanders are victims, they need victimizers. The pall of failure that hangs over the island, so cruelly demonstrated by Huntingdon's (1945) famous comparison of backward inbred Newfoundland with progressive vibrant Iceland, must be justified by external oppression. There are various candidates for the role of oppressor. The least convincing, and rarely used, is the Imperial power. London's almost total indifference for 500 years to the 'oldest colony' confined it to providing free external defence and, in 1932, a financial 'bail out'. Canada's use of the perpetual enemy, the United States, to support its fragile unity is also irrelevant here. Indeed the 'blame the feds' argument made possible by the reluctant entry into Confederation in 1949 is much commoner. 'Come near at your peril, Canadian wolf' is a widely expressed anti-confederation sentiment. However the most frequently encountered perpetrator of the island's woes is the home grown economic, political and social élite of St. John's. These are; the greedy merchants who fleeced the poor fishermen; the self-seeking political incompetents who bankrupted the Dominion, causing its return to colonial status (Matthews, 1973); and the culturally dominant 'townies' who patronized the simple 'baymen'.

The individual demonization of Premier Smallwood (1949-1972) as a result of his 'modernization' programmes combines most of these elements. The outport resettlement programmes have become the focus for all these grievances, to be nurtured and repeated with all the vehemence of a popular atrocity myth. The 'real' Newfoundlanders of the outports were betrayed by their own government into Confederation in 1949, their fish given away to foreigners and their settlements and way of life physically uprooted and destroyed, in return, as 'every bayman knows', for $600 a household. The visual image of the house being moved overland or across the bay has become a prevalent icon symbolizing the suffering of the bayman at the hands of an uncaring government.

The reaction of the oppressed to hard times produced the 'fighting Newfoundlander', resilient in the face of economic and political hardship as much as climate, topography and the sea. In part this is the image of the roistering sailor, expressing defiance of authority through shanties and *screech* (rum). In part it is the victim of circumstance who nevertheless bravely but futilely resists against hopeless odds. The two dominant iconic statues on the island are the caribou at bay and the soldier *en garde*, most movingly combined in the memorial to the Royal Newfoundland Regiment at Beaumont Hamel on the Somme.

The First People

The use of the past through history is related to the idea of folk, as both depend upon the valuation of antiquity, but is not identical with it. Folk is 'timeless tradition', the vernacular of the anonymous and unrecorded 'common people', whereas history is dated and chronicled events and personalities. The former is inward looking; the latter is used quite specifically to place Newfoundland in an external historical or geographical context.

Newfoundland is strongly promoted through the projection of the idea of 'oldest' and 'first'. The claim to primacy is unrelenting (Province of Newfoundland and Labrador, 1996). It is 'Britain's oldest colony' while St. John's is the 'oldest city in North America'. Numerous settlements claim 'firsts': official European Landfall (Bonavista); chartered settlement (Cupids); trans-Atlantic radio message (St. John's); trans-Atlantic cable (Hearts Content); and Atlantic flight (St. John's). This idea was central to the hallmark 1997 'Cabot 500', (as it was for the 1983 'Gilbert 400' and the 1897 'Cabot 400' before them). The theme has been continuously developed (Province of Newfoundland and Labrador, 1994; 1996) with the more recent additions of the Vikings (L'Anse aux Meadows as the first

European settlement in North America) and 'Leif Ericson 1000' in 2000. The first Basques, Irish (Brendan), Welsh (Caradog) remain a so far unexploited resource. Recently and somewhat tentatively the Beothuks been added as the first 'red' (that is, ochre-daubed) Indians and less happily, maybe the first to die out in the European era.

The Outport in Imagination and Reality

The elements of place identity in their various combinations must now be confronted with some of the functional realities, policies and aspirations of contemporary Newfoundland. All the senses of place outlined above are dependent in various ways upon the idea of the outport. This mundane adaptation to economic and physical circumstances combines and expresses Newfoundlandness as container, stage, artefact and symbol. A comparison of this outport of the imagination with the outport existing today raises wide discrepancies and dissonances (Tunbridge and Ashworth, 1997). The number, function, form and locational distribution of outports has changed and is continuing to change radically which in turn raises questions about the tenability, effectiveness and marketability of the image.

Most fundamentally they are fast disappearing as, simply, they no longer fulfil the functions for which they were created and new functions do not exist. The context within which the survival of the outport must be considered is what is euphemistically called the 'mobility option'. This is the official term used over the past 50 years for abandonment both locally (such as in the outport resettlement programmes) or even of, as has frequently been argued, the whole over-populated island (Overton, 1996, p. 3).

However changes in the number and location of outports is neither exclusively a recent nor a sudden phenomenon. There was no full and final permanent pattern of settlement which suddenly collapsed in the face of contemporary trends. The nature of fishing, lumbering, or quarrying communities imposes a certain impermanence leading to movement and even abandonment in response to changes in the resource-base and the demand for it. Government induced change only accelerated and possibly channelled such trends. The coast of Newfoundland is littered with abandoned outports and probably most had disappeared not just before the catastrophic decline in cod in the 1990s but also before the outport had been self-consciously labelled and recognized as a distinct and valued phenomenon. Three principal sorts of spatial change can be noted.

From Bays to Peninsulas

The outports were not as physically isolated from each other as later observers might imagine. Social and economic inter-outport networks were maintained by water transport as were periphery – centre links. Access to boats, and to the skills needed to use them, was effectively ubiquitous. Outports were therefore physically and socially grouped around the bays which defined the outport communities. On the northern shore this would have been Bonavista, Trinity, Conception, Notre Dame and White Bays, while on the southern shore it was Trepassey, St. Mary's, Placentia, Fortune and St. George's Bays. In this context island sites were at no particular disadvantage. The arrival of the road link to an island-wide road network, and the replacement of the boat with the motor vehicle switched the geographical focus from the bays to the peninsulas. This created a new pattern of accessibility which both created new links (for example, across peninsulas) and severed old ones (especially across the bays). Outports without road access, especially those on island sites were specifically disadvantaged and often rapidly abandoned. Tourism marketing reinforced the new pattern by its promotion of peninsula 'trails' and such peninsula based mental maps permeated beyond the tourist imagination to influence the local geography.

From Headlands to Bases

The switch from water to land transport not only re-focussed community networks from the bays to the peninsulas but conferred accessibility advantages on the base of these peninsulas rather than the headland. Previously, sites well up the peninsula were more accessible to both a variety of fishing grounds and to the merchants coasting around the island. Examples would include Grand Banks on the Burin Peninsula, Bonavista and Catalina/Port Union on the Bonavista Peninsula and Perlican or Bay de Verde on the Baccalieu Peninsula. These were now disadvantaged compared to locations near the base of the peninsulas with better access to the Trans-Canada Highway (TCH). There was some evidence of this even before the arrival of the motor vehicle and the TCH in the 1960s, as the railway whose main line ran from St. John's to Port aux Basque around the heads of the bays and bases of the peninsula had only two branch lines (namely the Bonavista spur through Port Union and the Baccalieu spur through Carbonear).

However car accessibility encouraged two changes in the settlement pattern. Within many settlements the commercial focus shifted from the waterfront to inland sites. In small towns such as Bay Roberts, Harbour Grace or Carbonear there was a clear commercial abandonment of the coast in favour of new commercial sites along the highway. This has resulted not only in the abandonment of buildings and the shift to an almost exclusively residential function for the old waterfronts but also to a physical and functional segregation of the old town and the new. This separation of the historic areas, now the heritage areas, physically and functionally from the rest of the settlement has implications for heritage management. Secondly, new regional service centres arose dominated by the car accessible shopping mall and strip. Existing settlements expanded to take on the new regional function (such as Marystown for the Burin) or Clarenville on the TCH for the Bonavista Peninsula). In 1901, Clarenville had a population of 229 and Trinity 1216: in 1991 the figures were 4473 and 326 respectively.

From Smaller to Larger

The combination of increasing demands for a wider range of services and the clear advantages to suppliers of a spatial concentration has encouraged the concentration of particular services and the clustering of diverse facilities. This has been the case for commercial providers (shops and other commercial services), governments (education and primary health care provision, post offices and emergency services) and non-profit agencies (particularly the religious service providers). The result, as in many rural areas, has been the relative growth of places such as Harbour Breton, Grand Bank, or Bonavista compared with the smaller outports in the catchment areas that they now serve.

Pasts and Futures: Tourists and Locals

In short the spatial patterns of functions, settlement and behaviour no longer concur with the image: the mental map no longer relates to the topographic map. Nowhere is the dilemma more clearly expressed than in tourism. The tourism networks, or promoted 'trails' are focussed on the peninsulas and not on the bays (Province of Newfoundland and Labrador, 1994). In addition tourism requires the provision of tourism services and road access. This by logical definition causes the outport to cease to exist. If the defining characteristic of an outport is physical isolation by land while the activation of the resource for tourism depends upon such access, then the presence of the tourist is not only a threat to the continuing existence of the resource, it is a sign that the resource, *sensu stricto*, no longer exists.

The outport is a combination of the visibly picturesque and the socially quaint. This in itself creates a number of ambivalences, not least in attempts to relate the past to the future and confront the intrinsic contradictions between a backward looking heritagization and a forward looking modernization. Although the projection of the image of wild nature inhabited by simple authentic folk has many advantages for both internal cohesive self-awareness and external recognition, it equally has numerous disadvantages that cause it to conflict with other images. A spiritually uplifting wilderness becomes a transport cost and an enriching solitude becomes an inconvenient isolation. 'Quaint' and 'traditional' can easily become 'backward' and 'old fashioned': a steadfast maintenance of time-honoured ways becomes a stubborn incapacity to embrace the present. This may result in little more than local irritation at the 'dumb Newfie' appellation but can have more directly serious consequences. It can serve to conceal and disguise problems which might otherwise be confronted:

> In trying to escape the capitalist consumer bubble and the world of staged authenticity and mass tourism, many seek "simpler" lifestyles and "real" culture in places like Newfoundland. What exists in such "backwaters",

however, is poverty, unemployment, inequality and desperation (Overton, 1996, p. 168).

It can also help frustrate policies to correct this situation. The dependence of the image upon the 'poor us' portrayal of Newfoundlanders as victims of natural misfortune, government ineptitude and a general failure to manage its economic, or political circumstances, is, to say the least, not a promotional element attractive to external investment.

The Smallwood administrations, for example supported the development of tourism as one part of a 'modernization from above' strategy. The promotion of the island for tourism would, it was hoped, support campaigns to attract multinational industry to the 'New' Newfoundland. Such policies also contained the idea that such an introduction of a 'modern' service industry would in itself 'modernize' the Newfoundlander. Tourism however commodified the very characteristics that it was expected to change and thereby reinforced them. The contradictions in this strategy are, at least with hindsight, obvious.

The physical support for the powerful idea of the outport is however sparse. The central dilemma concerns the question as to how a defunct settlement pattern can be commodified into heritage in support of both local and non-local place identities. The place image of Newfoundland, shaped over more than a century is based upon, and expressed through, forms that are technically difficult to preserve, unfeasible to re-use, and impracticable to visit. The solution is the creation of a number of compromises.

The Souvenir Outport

An obvious solution to the unvisitable outport is to consume its cultural resources as either folkcraft or nightlife elsewhere. In St. John's this tourism experience is spatially concentrated into the well-defined area of central Water Street and Duckworth Street, including the restored 'Murray Premises' speciality shopping Mall, and the interconnecting passages between the two. This area is now almost exclusively occupied by souvenir/craft retailing and food and drink facilities (Heritage Foundation of Newfoundland and Labrador, 1992; Mellin, 1999). The food and drink establishments stress combinations of elements drawn from the British 'pub', the waterfront seaman's tavern, the 'Celtic caile' and a newly discovered Newfoundland gastronomy (cod's tongues, seal flipper pie, moose stew and cod 'n brewis). The raucous sailor city is awash with 'screech' (rum), sea shanties and lusty sailors and their wenches in rowdy ale-houses' as promised by the St. John's Tourism Commission. All are simply marketing variations on the theme of the outport, its people, music, crafts, food, drink and visual image. The tourist can thus buy the product without the inconvenience of physical transport.

The Urban Outport

Another solution is the creation of the urban outport. The city of St. John's has expanded to include in its built-up area former outports which retain their physical form of buildings and narrow streets along the water (and are now scrupulously preserved as such). The best cases of these 'urban outports' are 'the Battery' overlooking the 'narrows' entry into St. John's Harbour and Quidi Vidi village. They provide the physical experience of the outport, although gentrification of these desirable properties excludes the social dimension of what otherwise is a virtual outport experience within strolling distance of the major hotels.

The Suburban Outport

A variation on the outport in the city, would be the suburban outport. Here settlements within daily commuting range of employment, particularly in the St. John's/ Mount Pearl city region (such as Torbay or Portugal Cove) become suburban settlements in which the idea, and possibly some surviving structures, of the outport is incorporated into the residential image. Similarly former outports at a greater distance from employment but still accessible in a few hours drive (especially those self-consciously pretty villages in physically attractive settings and architecture such as Brigus or Cupids) can take on a recreational function as part of the pleasure periphery of the city.

The Museum Outport

Given the central importance of the outport and its structures to Newfoundland identity, it is at first glance surprising how little attention has been formally paid to its preservation either *in situ* or in museums. The museumification of the outport has thus been largely left to local and often private initiatives. These can range from spectacular showpiece restorations and reconstructions (such as the rebuilding of the Ryan House at Trinity or the Kinsella premises at Fogo), small local museums run by voluntary societies (Heart's Content, Hibbs Hole), to little more than the exploitation of fortuitous abandonment rather than preservation (the steamer, 'Kyle', at Harbour Grace or railway station at Carbonear). The preservation of specific buildings is not of course preservation of an outport. For this area designation is required. Both Brigus and Harbour Grace are subject to heritage area protection (Heritage Foundation of Newfoundland and Labrador, 1992), the latter being chosen by the Heritage Foundation as the first test case for heritage area designation outside St. John's. Finally buildings and areas can the backdrop for the tourist-historic outport experience. Trinity has gone furthest in developing an outport experience for tourists and day visitors. The animated tourism trail and theatrical pageant in the Tidewater Theatre during the summer are a local initiative which has led to some local expansion of catering and overnight accommodation.

The Outport Heritage Trail

The recognition that a single outport is unlikely to detain the visitor longer than an hour, that accommodation is likely to be elsewhere (O'Dea, 1984) and that the car is the favoured means of transport, has led to the promotion of 'heritage-tourism routes' linking sets of outports. These generally have an historic, scenic or ethnic appellation ('The Captain Cook Drive', 'Marine Drive', 'The French Islands Drive', 'The Irish Loop Drive'). The fundamental difficulty of the 'heritage routes' is that being road orientated, they contradict the basic patterns of geographical accessibility that they are endeavouring to exploit. Traditional bay focussed regionalization is replaced by peninsula orientated heritage regions. The three most developed of these are all relatively close to St. John's and the eastern end of the island. The Avalon Peninsula, indented into what amounts to four separate peninsulas, is the most accessible and developed offering a number of circuits for day or weekend trips from St. John's that includes the major sites of Argentia, Placentia, Ferryland and Trepassey. The Baccalieu contains some of the most self-consciously pretty villages such as Brigus, Cupids, Carbonear or Harbour Grace. The Bonavista Peninsula is less accessible but has the notable towns of Trinity and Bonavista and proximity to the most accessible National Park on the island selling a combination of maritime history, whales and scenic interiors. The Burin Peninsula is the least accessible and frequented despite the potential attractions of Grand Bank, the south shore ferry at Bay L'Argent and the ferry for the French islands at Fortune.

The Developed Outport

St. Pierre has a similar physical setting and settlement history to other outports with the exception of the single distinguishing characteristic of being the last political remnant of French North America. This alone has determined that its development is quite unique. The concentration of national government investment, together with its Gallic characteristics has resulted in few of the outport attributes remaining extant. It has three main tourism products: the excursion potential of the 'island off an island' phenomenon, the attraction to 'collectors' of countries, and most important the selling of a 'little bit of France' in the New World to neighbouring markets, combining foreign exoticism with proximity and North American cultural familiarity. Such a development path is just not an option for the rest.

Conclusion

The outport of the imagination is thus unsellable not only to tourists (see Ashworth, 1998) but to almost any group that might be attracted to use it. The basic contradiction is simple. The 'frail, disordered trickles of settlement' (Neary and O'Flaherty, 1983, p. 10), characterized by the outport has already passed into

190   *Senses of Place: Senses of Time*

history. Its economic justification no longer exists, nor does the peculiar society with its distinctive cultural expressions, that it produced. However the symbolic significance of the outport as iconic representation of a valued idea is more important now for local consciousness and external image than it has ever been. It is *the* imagined Newfoundland. What had been originally projected externally has now been incorporated through reflection in the identity of Newfoundlanders. It has proved to be a creative resource supporting a highly flexible product range which can be sold to both residents, legitimating their place identities and to visitors as a defined brand-image, an on-site experience and purchasable souvenir. It is, in these respects, a relatively problem free heritage product range. The difficulties lie mostly in rendering preservable, promotable and accessible what is by its nature unpreservable, non-promotable and inaccessible. The heritage of the outport echoes so much of the bonanza history of Newfoundland in general. A potentially rich resource is combined with near insurmountable difficulties of its exploitation and transformation into tangible economic and cultural gains for the inhabitants of the 'Rock'. Like Cabot's ship, the 'Matthew', at Bonavista, Newfoundland as a whole has become a replica of a replica.

References

Ashworth, G.J. (1998), 'The Newfoundland Outport: The Unsaleable Tourism Product', *Trident*, Newfoundland Historic Trust, St. John's, **Spring**.
City of St. John's (undated), *Tourist Map*, Department of Tourism, Culture and Recreation, St.. John's.
Cohen, E. (1979), 'A Phenomenology of Tourist Experiences', *Sociology*, **13**, pp. 179-209.
Heritage Foundation of Newfoundland and Labrador (1992), *Harbour Grace: Heritage District Report*, St. John's.
Huntingdon, H. (1945), *Mainsprings of Civilization*, Wiley, New York.
Kurlansky, M. (1997), *Cod: A Biography of the Fish That Changed the World*, Walker, New York.
Matthews, K. (1973), *Lectures on the History of Newfoundland*, Maritime History Group, Memorial University of Newfoundland, St. John's.
Mellin, R. (1999), 'Fogo Island: Last Chance for Authentic Outport Culture', *Heritage*, pp. 10-13.
Mowat, F. (1972), *A Whale for the Killing*, McClelland and Stewart, Toronto.
Mowat, F. (1989), *The New Founde Land* , McClelland and Stewart, Toronto.
Neary, P. and O'Flaherty, P. (1983), *Part of the Main: An Illustrated History of Newfoundland and Labrador*, Breakwater, St. John's.
O'Dea, S. (1984), *Architectural Heritage of the Newfoundland Outport: A Preservation Development Strategy Based on Tourism*, MUN, St. John's.
Overton, J. (1980), 'Promoting the Real Newfoundland: Culture as Tourist Commodity', *Studies in Political Economy*, **4**, pp. 115-37.
Overton, J. (1988), 'A Newfoundland Culture?', *Journal of Canadian Studies*, **23**(1), pp. 5-22.
Overton, J. (1996), *Making a World of a Difference: Essays on Tourism, Culture and*

Development in Newfoundland, ISER, MUN, St. John's.

Province of Newfoundland and Labrador (1994), *A Vision for Tourism in Newfoundland in the Twenty-first Century*, Department of Tourism, Culture and Recreation, St. John's.

Province of Newfoundland and Labrador (1996), *500th Anniversary Travel Guide*, Department of Tourism, Culture and Recreation, St. John's.

Stuckless, A. (1986), *The Tourism Industry in Newfoundland and Labrador*, The Queen's Printer, St. John's.

Tunbridge, J.E. and Ashworth, G.J. (1996), *Dissonant Heritage: Managing the Past as a Resource in Conflict*, Wiley, Chichester.

Chapter 14

Media Production of Rural Identities

Peter Groote and Tialda Haartsen

Introduction

Local and regional identities are increasingly discussed in both academic and policy debates. It is vital to understand how place identities are formed and how localities and their histories play a role in this process. For academics, at least, identities are seen as social constructs that are continuously contested between actors. As identities deal with elusive concepts, such as meanings and values, actors produce and reproduce representations of places in this process of contestation.

This chapter focuses on representations of the rural. Research into rural representations normally considers the present-day situation. This may result in knowledge of current senses of rural places. But it provides little information on the temporal dynamics of the representations, and therefore on the mechanisms that created the current representations. This limits the possibilities of forecasting future representations. This chapter explores the development of rural representations over time by analysing the contents of a television programme on rural areas in the Netherlands.

Popular Discourses and Representations of the Rural

Nowadays, the focal point of mainstream rural studies is the analysis of the meanings that are ascribed to rural spaces. Understanding developments in rural areas must by definition start by interpreting such meanings and the representations of the rural that specific groups of actors hold. Examples of research into rural representations abound in recent issues of the leading journals (see, for example, Rigg and Ritchie, 2002; Saugeres, 2002; Van Dam et al., 2002). These and other analyses normally focus on specific groups of actors, such as farmers, urban and rural households (Halfacree, 1995), children (Jones, 1997; Matthews et al., 2000; Robertson, 2000; McCormack, 2002) or rural women (Little and Austin, 1996). Often such groups of actors have developed specific networks and practices of communicating meanings, and can accordingly be assumed to form separate discourses.

Discourses can be defined as 'specific series of representations, practices and performances through which meanings are produced, connected into networks and legitimised' (Johnston et al., 2000, p. 180). Although they should not be thought of as being independent, it is possible to distinguish between four separate discourses in rural studies: the lay, the academic, the professional and the popular. Halfacree (1993) is the most influential commentator in showing the importance of analysing rural representations within discourse analysis. He defined lay discourses of the rural as people's everyday interpretations and constructions of the concept of rural and of the places they consider being rural. In the lay discourse, representations are based upon communications within an individual's personal network. Halfacree contrasts lay with academic discourses, the latter being 'the constructs of academics attempting to understand, explain and manipulate the social world' (Halfacree, 1993, p. 31).

Jones (1995) adds the professional discourse and the popular discourse to the lay and the academic. The professional discourse is related to representations of people who are professionally engaged with rural areas, for example farmers or policy makers. Popular discourses, finally, are founded in cultural and societal structures, such as the media and fine arts. They differ from the lay discourse in that they extend beyond an individual's communicative network, 'to reach a wider public at a variety of scales ranging from local to international' (Jones, 1995, p. 38). The popular discourse is necessarily more strictly structured and organized than the lay discourse.

The Popular Discourse

The popular discourse often functions as an intermediary between the lay and the professional discourses. It both reproduces and produces the representations residing with the general public. On the one hand, mass media try to pick up and conform to the ideas and taste of the general public, as their ultimate goal is to 'sell' their messages in a competitive market to as large an audience as possible. On the other hand, by doing so mass media are very influential in shaping these very ideas and tastes: 'The media industries have been assigned a leading role in the cultural community of Europe: they are supposed to articulate the 'deep solidarity' of our collective consciousness and our common culture' (Morley and Robins, 1995, p. 174). Among the media in the popular discourse that have attracted geographers' attention are: photographs (Lutz and Collins, 1993; Dietvorst, 2002); movies (Burgess, 1994); tourism promotion flyers (Hopkins, 1998); or landscape painting (Howard, 1991). But since the 1960s the outstanding medium has undoubtedly been television. Leading Phillips et al. (2001, pp. 2-4) argue that 'rural geographers should be interested in the discourses of rurality at work in ... television.'

In the first place, television is now the dominant site of contemporary cultural production. The adage of 'the global village' originally referred to a TV-based society:

> It is now over thirty years since Marshall McLuhan became a media prophet by declaring that what made a TV-based society remarkable was not the content of the programmes but their mode of delivery. The medium is the message. What he noted was the instantaneity and ubiquity of information provided by television. ... The speed and quantity were similar to how news might spread in a small community yet the range of TV meant that this could include the whole world. Hence he suggested we were now entering the era of a global village (Crang, 1998, p. 94).

Magoc (1991) displayed the importance of television series in producing place identities in his analysis of the importance of five rural-based TV-series from 'The Real McCoys' to 'Lassie'. It made him conclude that:

> Rural-based television ... carries a multiplicity of messages, reflecting the complex and vigorous societal debate on the agricultural and environmental issues with which they often deal (Magoc, 1991, p. 25).

In the second place, it is clear that TV-broadcasted ruralities have become important symbols of specific rural localities. Morley and Robins's (1995), p. 10) comments on the construction of national identities seem as valid for rural identities:

> Broadcasting should also help to construct a sense of national unity. ... In the post-war years, it was television that became the central mechanism for constructing this collective life and culture of the nation (Morley and Robins, 1995, p. 10).

Again, it is telling that well over 20 years after 'All Creatures Great and Small' was first broadcast, Yorkshire is still sold on the tourist market as 'Herriot Country'. A more recent example is the boom in tourism experienced in the rural area around Bastogne (Belgium) since the broadcasting of the TV-series 'Band of Brothers'.

In the third place, TV is not just an important medium in shaping rural representations; it is also a well-documented one. This provides the researcher with the source material for a dynamic content analysis, tracing the evolution of elements of content over time. To quote Magoc (1991, p. 31) again:

> ... the rhetorics of television past can be used to trace the evolution of national discourse on social issues ...then we can chart the metamorphosis of modern American environmental attitudes in deconstructing the evolution of rural based TV.

In research into rural representations in the lay discourse, researchers have often turned to data collected through questionnaires or interviews (Halfacree, 1995; Van Dam et al., 2002; Saugeres, 2002; Haartsen et al., 2003a). From these, it is possible to assess which ideas people hold and which meanings they ascribe to the rural. These, however, are examples of static or diachronous research. Only longitudinal data collection really allows an analysis of changes in rural representations of specific groups with these methods. Haartsen et al. (2003a) suggested an alternative to longitudinal research, by translating current age-specific representations into the past and assuming that representations are to a large degree shaped in the formative years of each individual, running from the age of five to fifteen. Another alternative is to apply the methods of oral history by interviewing people about the representations they (think they) held in the past. The use of archival material on long-running events, such as a TV-series, provides an alternative for tracking developments in rural representations in the academic, professional, and popular discourses.

Analysing Rural Representations in the Popular Discourse

When analysing textual or visual data (photographs, paintings, movies, or television programmes), one has to consider both how they have been produced and which meanings they may communicate to an audience. They can be analysed using content, textual or discourse analysis.

Content analysis involves the identification of issues and the interpretation of the contents of the data source which is assumed to be significant (Hannam, 2002). It is a quantitative technique that is used for measuring meanings by coding, counting and categorizing occurrences or elements of contents. Most often, it uses directly measurable (manifest) characteristics of the text or picture, such as the number of published lines or articles (Caljé, 1997), or the occurrence of specific elements (for example, women or black people) in visual data (Lutz and Collins 1993). An important basic assumption of content analysis is the supposedly positive relation between the frequency of a theme or element and its importance (Hannam, 2002).

A distinction is sometimes made between the coding of the manifest content and latent content, and the cultural and symbolic meanings beneath the surface elements of a text of picture (Lombard et al., 2002). The latter is sometimes referred to as textual analysis or semiotics (De Bruijn, 1999; Hannam, 2002). This type of analysis is more time consuming and less standardized than content analysis of manifest content. Still one step further is discourse analysis. This focuses not so much on the text or image itself, but on who has produced the text, in what context, for which audience and, in particular, for which goals. The objective of discourse analysis is the disentanglement of the institutional and power structures that generated the text (Hannam, 2002). A warning is required here. In 1952,

Berelson still felt comfortable enough to describe content analysis as: '... a research technique for the objective, systematic, and quantitative description of the manifest content of communication' (cited in Lombard et al., 2002, p. 588). Nowadays, however, researchers are aware that, particularly in the analysis of visual data, that there is no such thing as an objective description. In order to keep the systematic character of the method intact, the assessment of intercoder reliability is therefore important. Intercoder reliability is a measure of the extent to which independent judges make the same coding decisions in evaluating the characteristics of messages (Lombard et al., 2002, p. 587).

In this chapter, we use content analysis as a research method for unravelling rural representations in a TV series. Methodologically we follow closely what Lutz and Collins (1993) have done in their seminal book on the cultural meanings of the post-war photography of *National Geographic Magazine*. They coded the contents of the visual data used, which were some 600 randomly selected photographs from the 1950 to 1986 editions. For each photograph, Lutz and Collins completed a coding frame that consisted of questions regarding 22 aspects. Among these were: the 'world location of the photograph'; 'smiling in photograph'; 'age of adults depicted'; 'urban versus rural setting'; 'skin colour'; and 'male' or 'female nudity' (Lutz and Collins, 1993, p. 285). Such coding may seem to devalue the rich and intricate nature of visual data but:

> Quantification does not preclude or substitute for qualitative analysis of the pictures. It does allow, however, discovery of patterns that are too subtle to be visible on casual inspection and protection against an unconscious search through the magazine for only those which confirm one's initial sense of what the photos say or do (Lutz and Collins, 1993, p. 89).

We applied a three-dimensional coding frame, 'a set of themes into which material can be allocated' (Hannam, 2002, p. 191) developed for earlier work (Haartsen et al., 2000, 2003a, 2003b). The methodology was initially developed for the analysis of data collected through a questionnaire. In this, respondents were asked to give four associations with the rural. These were classified into three orthogonal dimensions: the 'image base' dimension; the functional dimension; and the valuation dimension. The third dimension needs specific validation methods that are not available for our current data material.

The image base dimension assesses the fundamentals of each rural representation and is related to the way in which people form representations. Each association is classified in one of three discrete categories that together shape the image base dimension. Following mainstream cultural geography, Haartsen et al. (2003a; 2003b), following Foote et al. (1994), postulated the existence of three possible categories: a socio-economic functional image base ('how rural areas work'); a visual-figurative image base ('what rural areas look like'); a socio-cultural image base ('what rural areas mean'). The second, or land use, dimension is subdivided into five categories: agricultural uses; nature; residential; recreation; and infrastructure.

These two dimensions seem applicable to our current analysis as well, so that a comparison of the results with earlier research on rural representations in the lay discourse is possible. Application of the existing coding frame has the added advantage that intercoder reliability of keyword coding has already been assessed and reported (Haartsen, 2002, pp. 85-6). There is, however, a difference between coding keywords that respondents to a questionnaire have given when asked for words they associate with the rural on the one hand, and coding 10-15 minute items from a TV-documentary series.

Sources

Our data set with longitudinal data of representations of the rural results from a content analysis of a long run Dutch documentary TV-series, '*Van Gewest tot Gewest*' ('VGTG', meaning 'county by county') (see Haartsen et al., 2000). Between 1965 –2002, 'VGTG' was broadcast on a weekly basis on one of the two, later three, Dutch public broadcasting networks. It was targeted at a general audience which was supposed to be interested in developments in rural areas or, more generally, the periphery. In 'VGTG', each of the twelve provinces of the Netherlands was represented by its own editor, who also resided in his or her province. In September 2002, 'VGTG' was merged with its more urban focussed counterpart, '*Urbania*', into '*Gewest*' ('County'). Since the introduction of Dutch-language commercial broadcasting stations in the Netherlands in 1989, viewership figures had dropped. In 1988, the market share of 'VGTG' was about 30 per cent of total viewing time. Consequently, 'VGTG' may have been relatively important in influencing rural representations, at least until 1989. The archives of the Dutch Institute for Visual and Audio Data in Hilversum contain textual information on each broadcast of 'VGTG'. From this a database was built for each item broadcast between 1965 and 1992. The format of the catalogue changed after 1992, making further longitudinal difficult. The database contains information on the date of a broadcast, the region or locality, and the key words describing the item.

Results

Figure 14.1 shows the results of the content analysis of the TV-series, with regard to the image base dimension. From this it can be concluded that the image base remains remarkably stable in time. The shares recorded by the functional and socio-cultural image bases in the representations are larger than that of the visual-figurative one. This contradicts the results from the research discussed above into rural representations in the lay discourse (Haartsen, 2002; Haartsen et al., 2003a). According to these, the visual-figurative image base dominates rural representations and the functional image base in particular is less important.

From the relatively unimportant role of visual-figurative aspects in rural representations in 'VGTG' and the lack of dynamics in time, it can be deduced that the prevailing idea of the emergence of a visualized society with the increasing role of television in Western societies (Dietvorst, 1997) cannot be confirmed. It has to be stressed, however, that the visual-figurative part of the image base is not related to specific pictures of rural areas, but instead to associations with rural areas that relate to the way the countryside looks.

The share of the socio-cultural image base is more important in the rural representations in the TV-series 'VGTG' than in the representations in the lay discourse. However, socio-cultural aspects of the rural are not so important to substantiate the suggestion that 'television is often thought of as a source of an idyllic and mythical [...] view of the countryside' (Phillips et al., 2001, p. 4).

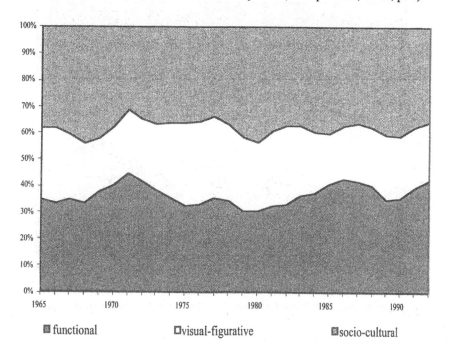

■ functional   ◻visual-figurative   ▨socio-cultural

Figure 14.1 Image base dimension in TV-series, 'VGTG'

Three remarks can be made on the shares of the different categories of the land use dimensions in rural representations, shown in Figure 14.2. First, there is a remarkable rise to prominence of nature and landscape as dominant functions in the way the countryside is represented on television. The increase in the number of nature and landscape items can be associated with the 1972 *Report of the Club*

of Rome, describing the limits to economic growth (Meadows et al. 1972). This
helped put the environment, including nature and landscape, on the political and
societal agendas. Secondly, a decline in the share of recreational land use in 'VGTG'
can be detected. This contradicts the booming rise of mass-tourism since the 1970s
and the subtler rise of 'quality' tourism at the end of the 1980s, for which rural
areas form a favourite context. Thirdly, agricultural land use only takes into account
a relatively small share of the rural representations in 'VGTG'. This does not
correspond with the results from the lay discourse, in which associations related
to agriculture form about 60 per cent of the representations of the countryside.
The share of residential land use is relatively large in the representations of 'VGTG',
compared to the lay representations.

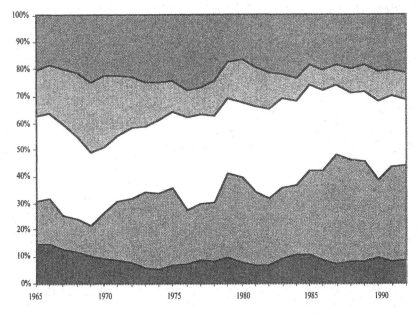

■ agriculture ▣ nature & landscape ☐ residential use ▨ recreation ▨ infrastructure & transport

Figure 14.2 Land use dimension in TV-series, 'VGTG'

Conclusion

The analysis of the contents of the television programme, '*Van Gewest tot Gewest*',
provides an introduction into the analysis of rural identities on Dutch television
and into the dynamic processes of the construction of senses of rural places in the
popular discourse. Rural representations in 'VGTG' are mainly formed along the

functional and socio-cultural image base. Residential land use is relatively important in the representations, while agriculture is underrepresented.

Apart from an increase in interest in nature and landscape at the cost of recreation, the rural representations in 'VGTG' are remarkably stable over time. This is surprising in the light of popular ideas regarding the importance of TV in expressing social debates. Magoc (1991), for example, stressed the importance of television series in producing place identities in his analysis of the importance of five rural-based TV-series, from 'The Real McCoys' to 'Lassie'. It made him conclude that:

> Rural-based television ... carries a multiplicity of messages, reflecting the complex and vigorous societal debate on the agricultural and environmental issues with which they often deal (Magoc, 1991, p. 25).

This may have to do with methodological problems in measuring exactly what we want and problems in the data as in, for example, the wrong key words being given to items by the library staff in the TV-series analysis. It is also possible that the analysis of just one TV-series is not enough and that it would be necessary to explore more programmes and other mass media for getting an overall picture of the dynamics in popular discourses (see Hannam, 2002). However, it may also be the case that TV is not as volatile and important in producing and reproducing the dynamics in societal debates as is commonly thought and that the methodological flaws are probably on Magoc's (1991) side. This is confirmed by Luke and Collins's (1993) quantitative content analysis results on indicators of wealth in *National Geographic Magazine* photographs between 1950-86 which compare surprisingly well with the data in Figures 14.1 and 14.2.

References

Berelson, B. (1952), *Content Analysis in Communication Research*, Free Press, Glencoe.

Bruijn, J. de (1999), *De Spanning van Seksualiteit. Plezier en Gevaar in Jongerenbladen*, Het Spinhuis, Amsterdam.

Burgess, J. (1994), 'Filming the Fens: A Visual Interpretation of Regional Character', in K.E. Foote, P.J. Hugill, K. Mathewson and J.M. Smith (eds), *Re-reading Cultural Geography*, University of Texas Press, Austin, pp. 121-39.

Caljé, J.F. (1997), *Lezen Doet Vrezen? Berichtgeving Over Milieurisico's in Dagbladen en de Reacties van Lezers*, Eburon, Delft,.

Crang, M. (1998), *Cultural Geography*, Routledge, London.

Dam, F. van, Heins, S. and Elbersen, B.S. (2002), 'Lay Discourses of the Rural and Stated and Revealed Preferences for Rural Living: Some Evidence of the Existence of a Rural Idyll in the Netherlands', *Journal of Rural Studies*, **18**(4), pp. 461-76.

Dietvorst, A. (1997), 'De Tijd-ruimtelijke Onteigening van het Toeristische Landschap', *Vrije Tijd en Samenleving*, **15**(2), pp. 5-14.

Dietvorst, A. (2002), 'De Illusie Gefotografeerd', *Geografie*, **11**(6), pp. 6-10.

Foote, K.E., Hugill, P.J., Mathewson, K. and Smith, J.M. (eds) (1994), *Re-reading Cultural Geography,* University of Texas Press, Austin.

Haartsen, T. (2002), *Platteland: Boerenland, Natuurterrein of Beleidsveld? Een Onderzoek naar Veranderingen in Functies, Eigendom en Representaties van het Nederlandse Platteland,* KNAG/FRWRUG, Utrecht/Groningen.

Haartsen, T., Groote, P. and Huigen, P.P.P. (eds) (2000), *Claiming Rural Identitie: Dynamics, Contexts, Policies,* Van Gorcum, Assen.

Haartsen, T., Groote, P. and Huigen, P.P.P. (2003a), 'Measuring Age Differentials in Representations of Rurality in The Netherlands', *Journal of Rural Studies,* **19**(2), pp. 245-52.

Haartsen, T., Huigen, P.P.P. and Groote, P. (2003b), 'Rural Areas in the Netherlands', *Tijdschrift voor Economische en Sociale Geografie,* **94**(1), pp. 129-36.

Halfacree, K.H. (1993), 'Locality and Social Representation: Space, Discourse and Alternative Definitions of the Rural', *Journal of Rural Studies,* **9**(1), pp. 23-37.

Halfacree, K.H. (1995), 'Talking about Rurality: Social Representations of the Rural as Expressed by Residents of Six English Parishes', *Journal of Rural Studies* **11**(1), pp. 1-20.

Hannam, K. (2002), *Coping with Archival and Textual Data,* in P. Shurmer-Smith (ed.), *Doing Cultural Geography,* Sage, London, pp. 75-92.

Hopkins, J. (1998), 'Signs of the Post-rural: Marketing Myths of a Symbolic Countryside', *Geografiska Annaler,* **80**(2), pp. 65-81.

Howard, P. (1991), *Landscapes: The Artist's Vision,* Routledge, London.

Ilbery, B. (ed.) (1998), *The Geography of Rural Change,* Longman, Harlow.

Jones, O. (1995), 'Lay Discourses of the Rural: Developments and Implications for Rural Studies', *Journal of Rural Studies,* **11**(1), pp. 35-49.

Johnston, R.J., Gregory, D. and Smith, D.M. (eds) (2000), *The Dictionary of Human Geography,* (Fourth edition), Blackwell, Oxford.

Little, J. and Austin , P. (1996), 'Women and the Rural Idyll', *Journal of Rural Studies,* **12**(2), pp. 101-11.

Lombard, M., Snyder-Duch, J., and Bracken, C.C. (2002), 'Content Analysis in Mass Communication: Assessment and Reporting of Intercoder Reliability', *Human Communication Research,* **28**(4), pp. 587-604.

Lutz, C.A and Collins, J. L. (1993), *Reading National Geographic,* University of Chicago Press, Chicago/London.

Magoc, C.J. (1991), 'The Machine in the Wasteland', *Journal of Popular Film and Television,* **19**(1), pp. 25-34.

Matthews, H., Taylor, M., Sherwood, K., Tucker, F. and Limb, M. (2000), 'Growing-up in the Countryside: Children and the Rural Idyll', *Journal of Rural Studies,* **16**(2), pp. 141-53.

McCormack, J. (2002), 'Children's Understandings of Rurality: Exploring the Interrelationship Between Experience and Understanding', *Journal of Rural Studies,* **18**(2), pp. 193-207.

Meadows, D.H., Meadows, D.L., Randers, J. and Behrens, W.W. (1972), *The Limits to Growth,* Routledge, London.

Morley, D. and Robins, K. (1995), *Spaces of Identity: Global Media, Electronic Landscapes and Cultural Boundaries,* Routledge, London.

Phillips, M., Fish, R. and Agg, J. (2001), 'Putting Together Ruralities: Towards a Symbolic

Analysis of Rurality in the British Mass Media', *Journal of Rural Studies*, **17**(1), pp. 1-27.

Rigg, J. and M. Ritchie (2002), 'Production, Consumption and Imagination in Rural Thailand', *Journal of Rural Studies*, **18**(4), pp. 359-71.

Robertson, M.E. (2000), 'Young People Speak About the Landscape', *Geography*, **85**(1), pp. 24-36.

Saugeres, L. (2002), 'The Cultural Representation of the Farming Landscape: Masculinity, Power and Nature', *Journal of Rural Studies*, 18(4), pp. 373-84.

Chapter 15

The Creation of Identities by Government Designation: A Case Study of the Korreweg District, Groningen, NL

M.J. Kuipers

Introduction

Since June 1994, building complexes and neighbourhoods built for social housing in the period 1910-1940 have been selected in The Netherlands, and put on the list of national monuments or national urban conservation areas (UCAs), as 'being in preparation for designation', to be eventually officially designated by the national government. In general, these social housing residential houses are characterized by some intrinsic weaknesses, they are for example built for a lower income residential population and in an experimental way with new building techniques and materials, that has led to most of them being, small (in both number of rooms and square meters), humid, and noisy.

The interesting point here is that it seems that history repeats itself. Similar social housing complexes were designed in the inter-war period by the cultural elite, that is urban designers and architects, that were convinced that they knew in what kind of house the working class population should (like to) live in, are now being selected for designation, or already being designated, by a same sort of paternalistic group of intellectuals that seems to be convinced they know that designating these social housing complexes or neighbourhoods as national monuments or UCAs respectively, is self-evidently beneficial in increasing the national stock of valued buildings, regardless of the demand for their historicity among the residents themselves. By means of a case study the valuation of the Dutch national and the Groninger local government of the younger UCA of the Korreweg district in the city of Groningen, built in the period 1910-1940 which is characterized by much social housing, will be compared with the valuation of the residents.

The Relation Between Historicity, Heritage, Identity, and Place

According to Ashworth and Howard (1999, p. 11) heritage is 'whatever people want to conserve, preserve, protect or collect' usually with a view to passing it on to others, and is best considered as a process which happens to things or as a marketing device. Heritage 'assumes a demand oriented approach in which the questions of selection are answered in terms of the demands of the consumer not the nature of the object being preserved' (Ashworth, 1991, p. 3). This means that selection criteria with respect to the object should not so much be based on its intrinsic qualities, such as age, beauty, or historical importance, but more on the needs of the market that are extrinsic to the object. This means that with respect to designation of an object, the focus should be on fulfilling this demand, instead of expecting that certain objects contain values that spontaneously will attract maintenance, protection and a passing on to future generations. Graham et al., (2000, p. 2) define heritage as 'the contemporary use of the past, based on the present needs of people', and argue that from this point of view an object does not per se need to be 'historically correct, intrinsically authentic, innately valuable or qualitatively worthy'. 'If people in the present are the creators of heritage, and not merely passive transmitters or receivers of it, then the present creates the heritage it requires and manages it for a range of contemporary purposes' (Graham et al., 2000, p. 2). They interpret heritage as a concept of meaning rather than artefact, that ensures it is a field of social conflict and tension, carrying differing and incompatible meanings simultaneously.

Historicity means 'being historic', and historic means 'according to, originating from, going with the succession in time, or history' (Pijnenburg and Sterkenburg, 1984, p. 503). In other words historic means 'in one way or another related to the past'. Ashworth (1994a, p. 8) defines the past with respect to the planning of heritage as 'a collection of resources – symbols, associations, or relicts–selected and activated by the present for different kinds of use on behalf of a future, that is designed by spatial policy'. The past is dynamic, in that it 'creates the present, that in turn becomes the past very quickly' (Ashworth, 1994a, p. 1). Place-historicity is the extent and the way in which a place reminds people, or makes them think, of the past. Rose (1995, p. 88) argues that identity is 'how we make sense of ourselves' and that sometimes the meanings given to a place may be so strong that they become a central part of the identity of the people experiencing them:

> Although identity refers to lived experiences and all the subjective feelings associated with everyday consciousness, it also suggests that such experiences and feelings are embedded in wider sets of social relations: they are shaped in large part by the social, cultural and economic circumstances in which individuals find themselves' (Rose, 1995, pp. 88-89).

According to Ashworth and Kuipers (2001), identity is ascribed to landscapes not given by them. For this reason, the active variable is not the landscape, object or place, but the 'someone'. Two other points about place-identity that should be made clear here is first that it is plural not singular: 'different people at different times identify in different ways with places', and secondly that it is dynamic not static: 'cultural heritage is a reflection of what we have decided we want to reflect' (Ashworth and Kuipers, 2001, p. 59). According to Carter et *al.* (1993, p. xii) 'place is space to which meaning has been ascribed'.

Historicity and Heritage

History is 'the occurrences, or the remembered record, of the past' (Ashworth, 1994b, p. 13). In turn, historicity is the extent and the way in which a specific object reminds, or makes people think, of the past. Ashworth (1994b, p. 16) argues that history becomes heritage through a process of commodification, and defines heritage as 'a contemporary commodity purposefully created to satisfy contemporary consumption'. This means that place-history is subject to two valuation processes. First, in an active, conscious, and top-down way, namely the commodification of a place by its 'producers', based on their interpretation of history and the way they want to give the history-message through to the public. Secondly, the historicity of a place according to its 'consumers' in a passive, unconscious, and bottom-up way, is the extent to which the place reminds them or makes them think of the past, and the extent to which they appreciate, for example in their orientation with respect to, or identify with, a specific place. The point here is that the commodification of history by the heritage-producers does not always concur with the appreciation of history by the heritage-consumers, because of the fact that they use the history of a place for different reasons, and therefore value it differently, and in turn appreciate different aspects of place-specific history.

Heritage and Identity

Ennen (1999) argues that historic objects tell their users different stories that play a part within their cultural identity. As a reaction to an increasingly globalized culture it has become necessary and desirable to distinguish places from other places, in which historicity can play an important role (Ennen, 1999). What should be pointed out here, is that according to Ashworth (1994b) it is not history, as defined as the occurrences of the past, that is used as a socio-psychological, political, economic, or cultural resource, but heritage. According to, Graham et al. (2000) places are distinguished from each other by many attributes that contribute to their identity and to the identification of individuals and groups within them. Heritage can be seen as one of these attributes: 'the sense, or more usually senses of place is both an input and an output of the process of heritage creation' (p. 4). According to Ashworth (1994b) heritage is one of the main determinants of places, and one of the principal components of real differentiation. Several motives exist for the maintenance and protection of heritage, such as a political-administrative motive, a romantic-nostalgic motive, a scientific motive, or an economic motive (Kersten, 1998). All these motives have one thing in common and that is identity: 'it is not only about the building 'as such', but also about the story behind it, and about the extent in which it gives identity both to the building and the community' (Kersten, 1998, p. 454).

Valuation and Appreciation of Recent Heritage by its Experts and Users

Heritage-experts

In The Netherlands, the policy for the protection of monuments (Monumentenzorg) is institutionalized nationally in the National Service for the Care of Monuments (Rijksdienst voor de Monumentenzorg) since its establishment in 1875. Especially until 1960 it strongly focused on selection criteria of preservation planning, however since 1960 it still does but in a less extreme way. Within preservation planning the motivation for designating buildings, residential houses, residential districts, or inner cities is based on their supposed self-evident intrinsic qualities such as rarity, age, beauty, historical importance, architectural importance, or special architectural design (Ashworth, 1994a). Whether a building or area contains these qualities is decided by the so-called 'heritage-experts', such as architects, artists, urban designers, archaeologists, in a professional (Rijksdienst voor de Monumentenzorg) or in a more amateurish way.

Selection criteria of conservation planning, the movement that dominated in the period 1960-1976, but still exists, are based on the extrinsic functioning of the area, its goals are among others: strengthening the urban form, and the effective functioning of cities (Ashworth, 1994a). These extrinsic goals are not named specifically in the motivation of designation decisions of the Rijksdienst voor de Monumentenzorg, both in the case of national monuments and even of whole urban conservation areas, that form the main focus of conservation planning. Probably, because its sees the extrinsic functioning of the area designated as the concern of professional urban planners and local political groupings. With respect to this should be mentioned that in the definition of 'urban and rural conservation areas', as formulated in the Dutch Monuments Law of 1988, a number of criteria are mentioned, however none of them is based on its future use.

Heritage planning is especially a development of the 1980s, as a consequence of the success in conservation policy towards monuments and UCAs, and the problems that this success itself has created, such as rising financial burdens, and restrictions on the efficient functioning of cities as a consequence of the fossilization of their structure (Ashworth, 1994a). Its selection criteria are based on the needs of the market, that are totally extrinsic to the object, and its goal is based on the satisfaction of the recent or future needs of the market. In reality, the motives of the Rijksdienst voor de Monumentenzorg for designation are, in the first instance, not based on the demand for the object in question, rather (as mentioned earlier) on the intrinsic values it ascribes to it.

Heritage-users

According to Ashworth (1991, p. 70) the user of the historic city is only definable in terms of the individual intent of the user at the moment of use: 'an individual may successively work, shop, and recreate in the same area of the historic city, moving rapidly and often unconsciously between categories'. He argues that the relationship between users of the historic city and its historicity, in terms of their actual use, is a complex one. In 'Heritage planning' Ashworth (1991, pp. 70-1) describes several classifications of users of the historic city to show that they are multi-motivated. One classification consists of: (1) residents from the city-region, incidentally drawn to the historic city by its distinctive attributes, for which the historicity of the city is not their main motive; (2) residents from the city-region intentionally drawn to the historic city by its distinctive attributes, also called 'trippers', for which the historicity of the city is their main motive; (3) visitors from outside the city-region incidentally drawn to the historic city by its distinctive attributes, for which the historicity of the city is not their main motive; (4) and visitors from outside the city-region intentionally drawn to the historic city by its distinctive attributes, also called 'tourists', for which the historicity of the city is their main motive. However, although historicity is not a user's first motive for visiting the historic city it could be something they incidentally enjoy. In this respect, Ashworth (1991, p. 72) states that museums, historic and cultural attractions fulfil two different roles: 'they function as "primary" resources, i.e. actually serve as motives for the trip, although usually in association with other motives such as business, shopping or recreation: they in addition function as "secondary" resources, i.e. they provide an ancillary service or enhancement on trips quite differently motivated'.

In this way, also an historic residential house or district can be seen both as a 'primary' resource in the case its residents choose to live there primary because of its historicity. However when this will not be the case, the historic residential house or district can serve as a secondary resource in the sense that its historicity enables people in general to identify themselves with, or to orient themselves on the basis of, the historic residential house or district.

Valuation and Appreciation

There is a renewed interest in the quality of urban life, among cultural critics, planners, and property developers. It can also be argued that the post-modern city can be regarded as the antithesis of modernist abstraction and anomie, and that it is about the renaissance of urban culture and sensibilities. In 1994 the Dutch national government started the so-called Monuments Selection Process (MSP) in which objects, complexes and areas from the period 1850-1940 are selected. Its aim is the preservation of valuable heritage with respect to architectural history, cultural history and city planning on the basis of the Monuments Law of 1988. In 1999, more than 300 urban and rural conservation areas from before 1850 were officially designated in the Netherlands. As a consequence of the MSP, about 165 urban and rural areas from the period 1850-1940 will be added to the existing list (Nelissen, 1999). On the basis of these figures it can be seen that according to the experts, in this case the *Rijksdienst voor de Monumentenzorg*, these recent historic areas have a clear heritage value.

According to Kersten (1998) spatial planning is often about the choice between demolishing and preserving, especially in the case of the buildings from this century. In this choice the identity of the building plays a large role. 'When the identity is desirable and there is enough support for it, than the building will be preserved, when there are negative associations with a building, then usually it will be demolished' (Kersten, 1998, p. 455). What Kersten forgets to explain here is who decides what kind of identity is desirable or not.

On the basis of a research of Ganzeboom (1983) about the visiting and valuation of monuments in the city of Utrecht, Ennen (1999, p. 74) argues that: (1) users receive considerable pleasure from visiting monuments; (2) listed monuments are often more positively valued than non-listed monuments; (3) the more recently (after the 1860s) buildings have been built the less positively they are valued. However, Ennen (1999) stresses that residents often have a different, more passive and unconscious way of valuing their historical built environment than the visitors to the same environment.

Laenen (1989) argues that the extent to which cultural heritage, and the attainments and cultural values of the past will be integrated and continued, will not only depend on their inherent qualities, but also on their validity to present day society. In this respect, Kersten (1998) argues that within recent conceptions a reference to the past often guarantees a high valuation. However this raises the question if the same applies to the recent past.

The Relations in a Model

Figure 15.1 shows that every place, such as a whole city or just a residential district, is valued – in a positive, negative, or neutral way – both by heritage-experts and by heritage-users. The first group values a place more on political-administrative grounds, and the second more on socio-psychological grounds. As a consequence, the experts will more prescribe, and the users more ascribe, historicity to a certain place. The extent to which a place is able to become heritage, in other words its heritage-potential, depends on the extent to which it is valued as historic by both the experts and the users. In order to survive as heritage, a place should not only be valued as being historic, but should also be appreciated as such. This appreciation will influence the way in which the experts will prescribe, and the users will ascribe, identity to a place. In turn, the extent in which this takes place will influence its heritage status, which in turn will influence the extent in which a place again will be valued.

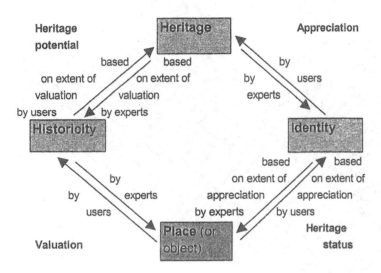

Figure 15.1 A model of the relations between place, historicity, heritage and identity

The Case Study

The UCA of the Korreweg district (see Figure 15.2), built in the period 1910-1940 in the city of Groningen, is an example of early social housing in The Netherlands. However it is not the only social housing district that already is, or still has to be, designated, as a younger UCA in the Netherlands. Other examples of residential districts originally and still characterized by a lot of social housing that already are or probably will be designated as younger UCAs are De Muntel, Geitenkamp, De Bomenbuurt, and Hilversum-Zuid (RDMZ, 2000). These are located in the following Dutch medium sized (as is the city of Groningen) cities: 's-Hertogenbosch, Arnhem, Haarlem, and Hilversum respectively. Also in other West European countries these examples of the early social housing residential districts can be found, such as the Mile Cross Conservation Area. This is a large residential area in the north west suburbs of the medium sized city of Norwich in the United Kingdom which was constructed during the inter war period by Norwich City Council for Social Housing (Insley, 2001).

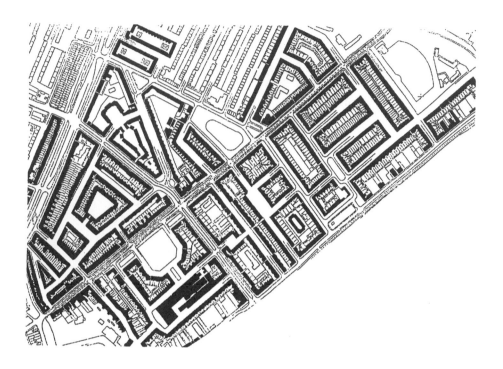

Figure 15.2 The Korreweg UCA

Source: *Rijksdienst voor de* Monumentenzorg (2003)

    Because every medium-sized or large city in The Netherlands will have one or more residential districts, that contains one or more early social housing complexes, and because this would probably also be the case in similar cities in the rest of Western Europe, it is important to find out what kind of heritage status these social housing residential districts have according to its residents on the one hand and the Rijksdienst voor de Monumentenzorg on the other. In other words, do the identities that the national government wants to create by the designation of these districts as a younger UCA concur with the identities ascribed to them by their actual users, namely their residents?

Valuation of the Korreweg UCA by the National and Local Government

In 2000, the Rijksdienst voor de Monumentenzorg designated the Korreweg district in the city of Groningen because of the following, rather general, values they 'prescribe' to it:

> (1) It is a complete example of an extensive district of the town from the 1920s and 1930s with a well preserved clear architecturally design, a spacious design and an architecturally coherent and varied built-up area, in which the different phases of development are clearly recognisable.
> (2) It possesses spatial quality, which is supported by the spacious profiles of the main streets, squares and ponds with varied green, and the spatial organization of the street-pattern. (3) It is of general importance, because of its meaning for the development in thinking in terms of urban design, the quality of the spatial concept, the scale and the architectural elaboration (Kuipers, 2002, p. 57).

In 1995, the *Rijksdienst voor de Monumentenzorg* designated 272 residential houses as national monuments, of which most form part of a larger residential complex, in the younger UCA of the Korreweg district (RDMZ, 1995). The decisions to designate these residential houses or complexes were based on the following, again rather general, values, named by the national government: (1) cultural-historical and architectural-historical value, (2) architectural, esthetic, building-technical quality of the design, (3) its meaning for the history of urban design and social housing, (4) its uniqueness and rarity (RDMZ, 1995).

Valuation of the Korreweg UCA by its Residents

In a case-study consisting of 310 questionnaires, the residents of residential houses not designated as a national monument located in the younger UCA of the Korreweg district were asked to what extent they found their residential district historic or not. For this purpose, several questions about the valuation of their residential district with respect to some indicators, namely 'age', 'specificity', 'value', 'beauty', 'character', 'meaning', 'atmosphere', and 'past', were asked them. These indicators, or historicity-aspects, are based on the several categories of answers on the question why they found their residential district and residential house historic, in an earlier carried out pilot-study among 130 respondents of the Korreweg district (see Kuipers, 2002, p. 59).

Approximately 91 per cent of the 310 respondents agreed with the designation of their residential district by the national government. A large part of the respondents (73 per cent) disagreed with the following statement: 'My residential district is not yet old enough to be designated as an urban conservation area'. From this it seems that most of the respondents found their residential district worthy of designation because of its age.

These results show that the historicity-aspects of the younger UCA of the Korreweg district are valued, that is recognized and acknowledged, by most of the respondents, in a moderate or neutral way. If all the historicity aspects are taken together then it seems that in general most of the residents find their residential district 'historic', however not 'very historic', and another notable part finds it somewhere in between 'historic' and 'non-historic'.

Appreciation of the Korreweg UCA by National and Local Government

Kloosterman (1994) argues that the designation of a residential district as a UCA will create some opportunities: (1) it can support and stimulate, already existing or to be developed, policy on spatial quality of the specific district, (2) it can serve as a framework for policy on image quality.

According to the Fourth National Policy Document on Spatial Planning (1988), the spatial quality of an area is defined by its 'functional value', 'value of experience', and 'value for the future' (Voogd, 2001, p. 25). These three values refer to the dimensions 'function', 'form', and 'time', respectively.

In the Fifth National Policy Document on Spatial Planning (VROM/RPD/ CDC, 2001, p. 21) 'spatial quality' can be achieved by preventing that space becomes more monotonous, in order to provide a residential environment in which people can live according to their wishes and needs, so that they are able to function in it, and to enjoy it. This means according to the report that cultural and historical values should be protected, and variety and contrast promoted. The report names seven criteria of spatial quality, of which 'spatial diversity', 'attractiveness', and 'human scale' are of interest here.

With respect to the first concept, the report describes that 'diverse urban environments and landscapes must be able to preserve and strengthen their 'own' character' (VROM/RPD/CDC, 2001, p. 23). With respect to the second concept it argues that 'maintaining urban attractiveness is a cultural duty, and that greater attention to design and planning is required for the city' (VROM/RPD/CDC, 2001, p. 23). Voogd (2001, p. 26) has a clearer description of what is meant by this criterion, namely 'preserving the beauty of landscapes and cities'. With respect to the third concept it argues that 'the way in which space is planned must fit citizens' demands and perceptions' (Kersten, 1998, p.455). 'Image quality' is according to Kersten (1998) shaped by both the residue, that is the renovation and protection of concrete heritage, and the identity, that is the renovation and protection of stories about and from the past.

Laenen (1989) argues that the function of the history and its expression in heritage is not only to assert and further develop cultural identity and to be social constructive in terms of it, but is also to contribute to the quality of our present-day life. Kersten (1998) names three categories of 'image quality' plans, based on three different levels of scale: (1) the 'whole city', which means that on city-level there will be looked at the identity-carriers of the city and that these will be emphasized or strengthened in a project-like way, (2) the 'existing urban sub-area', in which it is most of the time about the fitting in of new objects in the recent built environment, (3) the 'to-be-developed urban area', in which there has to be looked for connections with the identity of the rest of the city. 'Image quality' lies between 'orientation', often based on clearly structured situations within a city, and 'surprises', often based on variation and diversity, and the perfect instrument in the strengthening of these two, is the historic built environment (Kersten, 1998).

The local government of the city Groningen believes in the following positive knock-on effects as a consequence of the designation of the Korreweg district as a younger UCA.

'The designation of monuments, urban conservation areas ... contribute to the strengthening of the identity and quality of the residential district or neighbourhood', and: '... the policy for the care of monuments *(Monumentenzorg)* can be used as an instrument in the managing of the continuation of the development process in order to remain the valued image quality and characteristic resulting from that' (Dienst RO/EZ, 2000, pp. 52-3).

The same local government argues that some specific characteristics of the Korreweg-Oosterpark district (of which the Korreweg district forms a part), namely 'the monumental architecture', 'the clear spatial structure', and 'the favourable location with respect to the inner city' will make it an attractive urban residential environment (Dienst RO/EZ, 2001, p. 2).

Appreciation of the Korreweg UCA by its Residents

For the largest part (around 44 per cent) of the respondents, an old and characteristic appearance of a residential house is 'important' in their residential choice (see Figure 15.3). However a smaller but still relatively large part (26 per cent) found it somewhere in between 'important' and 'unimportant'. Although a significant part of the respondents said that an old and characteristic appearance of a residential district is important in their residential choice, Figure 15.4 shows that it is certainly not a primary motive in their choice of going to live there. This probably means that the historicity of the Korreweg district does not contribute a lot to the attractiveness of the residential district from the point of view of the residents. Location near the inner city seems to play a far more important role.

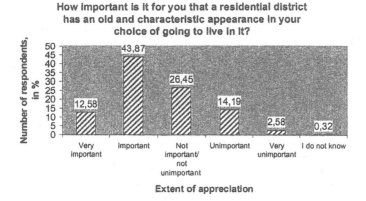

Figure 15.3 Appreciation of a residential district as being historic

Source: Kuipers (2002)

Conclusion

Most of the residents of the younger UCA of the Korreweg district value their residential district in a moderate or neutral way. Although the historicity of the district is to a certain extent valued by the residents, the question remains whether they do appreciate the historicity they ascribe to it. In fact, the extent to which they appreciate their ascribed historicity, seems to be low. This is based on the fact that a very small part of the respondents named historicity as their primary reason in their residential choice. This makes the extent in which the residents identify themselves with, or feel a sense of belonging to the residential district, because of its historicity, questionable.

The historicity of the Korreweg district is highly valued by the Dutch national and local government. Their appreciation of their prescribed historicity of the district is also high, because they believe that the historicity of the younger UCA of the Korreweg district (ultimately) will lead to a higher 'spatial and image quality' of the district.

On the basis of the valuation and the appreciation of the historicity of the younger UCA of the Korreweg district by its residents can be said that in their eyes it has a clear, but not a very clear, heritage potential and heritage status. When looking at the valuation and the appreciation of the historicity of the younger UCA of the Korreweg district by the experts, than can be said that according to them it has a clear, or even very clear, heritage potential and heritage status.

Which characteristic of your residential district played in that time
the most important role in your choice of going to live in it?

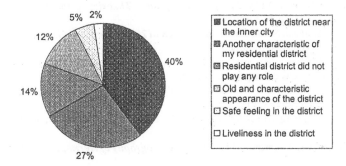

Figure 15.4 **Primary motive in residential choice**

Source: Kuipers (2002)

So, there is some difference in valuation and appreciation of the historicity of the Korreweg district between its users and experts. However it is not very large, and has a chance of becoming smaller because of two processes. On the one hand the intrinsic ageing of the specific built environment, and on the other the possibility of attracting residents that will appreciate its historicity.

Here, the main conclusion is that in creating identities by means of designating a younger UCA, the government should not ignore the actual identity ascribed to it by its residents. Before designation it is recommended that they first find out if its historicity is both valued and appreciated by its residents and to what extent, and secondly if it fits the way it is valued and appreciated by the national and local government. Because, if heritage is demand-based, so also should be its designation.

References

Ashworth, G.J. (1991), *Heritage Planning*, Geo Pers, Groningen.
Ashworth, G.J. (1994a), *De Pijl des Tijds in het Ruimtelijk Doel*, Geo Pers, Groningen.
Ashworth, G.J. (1994b), 'From History to Heritage: From Heritage to Identity: In Search of Concepts and Models', in G.J. Ashworth and P.J. Larkham (eds), *Building a New Heritage: Tourism, Culture, Identity in the New Europe*, Routledge, London, pp. 13-30.

Ashworth, G.J., and Howard, P. (1999), *European Heritage Planning and Management*, Intellect, Exeter.

Ashworth, G.J. and Kuipers, M.J. (2001), 'Conservation and Identity: a New Vision of Pasts and Futures in the Netherlands', *European Spatial Research and Policy*, **8** (2), pp. 55-65.

Carter, E., J. and Squires, D.J. (eds) (1993), *Space and Place: Theories of Identity and Location*, Lawrence and Wishart, London.

Dienst RO/EZ (2000), *De Stad van Straks Extra, Groningen in 2010, Gemeente Groningen*, afdeling Beleidsontwikkeling, Groningen.

Dienst RO/EZ (2001), Op Weg naar een Stadsdeelvisie voor Korreweg/Oosterpark, Gemeente Groningen, afdeling Stadsdeelcoördinatie, Groningen.

Ennen, E. (1999), *Heritage in Fragments: The Meaning of Pasts for City Centre Residents*, Nederlandse Geografisch Studies, 260, Groningen.

Ganzeboom, H. (1983), *Beleving van Monumenten (2)*, Sociologisch Instituut, RUU, Utrecht.

Graham, B., Ashworth, G.J. and Tunbridge, J.E. (2000), *A Geography of Heritage*, Arnold, London.

Insley, P. (2001), E-mail at 2 Nov 2001 from Phil Insley, Conservation and design officer at the local government of Norwich, in the United Kingdom.

Kersten (1998), 'Beelden van verleden', *Rooilijn*, **9**, pp. 453-58.

Kloosterman, J.O.D. (1994), *Enige Aspecten van de Aanwijzing als Beschermd Stadsgezicht*, *Zeist*, RDMZ, Regioservice.

Kloosterman, J.O.D. (2000), interview at 1 Dec 2000 with Okko Kloosterman, Senior Consultant 'Urban Design for Eastern and Northern Regions of the Netherlands', at the National service for the care of monuments in Zeist, The Netherlands.

Kuipers, M.J. (2002), 'Consequences of Designating the Recent Past: Korreweg-district, Groningen, The Netherlands', *International Journal of Heritage Studies*, **8**(1), pp. 53-62.

Laenen, M. (1989), 'Looking for the Future Through the Past', in D.L. Uzzel (ed.), *Heritage Interpretation: The Natural and Built Environment*, Belhaven Press, London, pp. 88-95.

Nelissen (1999), *Monumentenzorg in de Praktijk*, Nelissen/ NCM/NRF/RDMZ, Nijmegen.

Pijnenburg, W.J.J., and Sterkenburg P.G.J. (1984), *Van Dale: Groot Woordenboek van Hedendaags Nederlands*, Van Dale Lexicografie, Utrecht/Antwerpen.

RDMZ (1995), *Beschermd(e) Monument(en) Opgenomen in het Register Ingevolge Artikel 6 van de Monumentenwet 1988*, Rijksdienst voor de Monumentenzorg, Zeist.

RDMZ (2000), 'Lijst van Stads- en Dorpsgezichten uit de Periode 1850-1940 Waarvoor een Aanwijzingsprocedure in Voorbereiding is', *Rijksdienst voor de Mionumentenzorg*, Zeist, pp. 19-24.

Rose, G. (1995), 'Place and Identity: a Sense of Place', in D. Massey and P. Jess (eds), *A Place in the World?* Oxford University Press, Oxford, pp. 87-132.

Voogd, H. (2001), *Facetten van de Planologie*, Kluwer, Alphen aan den Rijn.

VROM/RPD/CDC (2000-2001), *Samenvatting Ruimte Maken, Ruimte Delen: Vijfde Nota over de Ruimtelijke Ordening 2000/2020*, Ministerie van Volkshuisvesting, Ruimtelijke Ordening en Milieubeheer, Rijksplanologische Dienst, Centrale Directie Communicatie, Den Haag.

CONCLUSIONS

Chapter 16

The Next Questions

The Editors

An interest in sense of place is as old as the study of geography itself and the idea that such identities are created and recreated by the actions of people is, as we have observed, widely accepted. What is new, however, as borne out by many of the preceding case studies, is the increasing interest of official government agencies at various levels in 'sense of place'. Quite inevitably, this further increases the potential for contestation between such official representations and unofficial narratives of place, often shaped as a conscious act of resistance against the state. Moreover, the institutionalization of 'sense of place' through heritage policies may also enhance the degree of dissonance that can exist between communal and individual perspectives on place and time. One such case, for example, is the national inter-ministry, long-term Dutch government policy programme known as 'Belvedere' which was initiated in 1999. This political strategy to link cultural history and spatial planning is more comprehensive and better financed than most other national heritage programmes and thus poses much more widely applicable questions that stem directly from the issues and concerns addressed in this book.

The objectives of the Belvedere programme are to locate, label and map all those landscape regions and cities in The Netherlands, which are perceived as having a clear, distinctive character and which, therefore, can contribute to the creation or enhancement of a local identity. This raises the immediate practical question of recognition. What is this local identity that is being sought and how can it be recognized? It should be stressed that there is no explicit mention in the policy documentation of any national stereotype of landscape or cityscape whose local manifestation is to be sought. It is assumed that the country is a palimpsest of localities which are defined by some common collective identity as a concept similar to that of collective memory. But collective identity poses the same question, as does collective memory: is this an aggregate summation of a myriad of individual identities or something quite separate and plausibly different? In addition, in The Netherlands, the under-riding assumption of the Belvedere programme is that a coherent national identity is shaped from the myriad local. This contrasts with unagreed societies where, as for example in Northern Ireland, the same palimpsest has no sense of commonality. Thus The Netherlands seems to suggest a 'Russian doll model' of comfortably nesting identities, ranging in size from the single individual to the largest collectivity applicable, whereas Northern Ireland suggests that models of conflict are more relevant.

In practical terms, the Belvedere programme depends very largely on local government agencies responding locally to the stimulation and opportunities offered by the national bodies. While heritage conservation is generally concerned with place as a collection of physical elements that can be physically or legally protected, it is much less clear how to protect a local community or, as in this case, a local identity created by such a community. Both communities and identities are in a process of constant change and are not static entities capable of being frozen at a particular moment in time

This points to several questions about the links between senses of time and place. In many ways, the discussion of identity echoes the parallel discussion on heritage as the contemporary uses of the past. Heritage draws upon elements of history, memory and selective relict artefacts as resources to effect a self-conscious anchoring of the present in a selected time context. This dominance of the past, however, raises the twin dangers of creating an identity based upon social and cultural elements that are already obsolete and largely irrelevant to the daily way of life of most locals, while also possibly fossilizing past or present patterns in a way that will inhibit future change.

In the context of the Belvedere programme, the motivation for what is a heavily subsidized government initiative is quite explicit, namely, the strengthening of local identities as a counterpoise to increasing economic and cultural globalization which is seen as threatening to produce a homogenous universal 'placelessness' that is assumed to be undesirable. This places the programme squarely in the much wider debate about the impacts of globalization/localization. There is a search for a 'balance', here assumed to be the realization of the economic gains of globalization while compensating in the cultural sphere by a support for localization or at least a mitigation of the losses anticipated locally. However, both sides of the balance can be questioned. Globalization may instigate or accelerate change in senses of place leading even to the much-feared death of locality but, equally, it may be only the substitution of one place-identity for another at a different scale. The 'global village' remains local in one sense if not another. Similarly the attempt of national governments to support a sense of local identity may of course lead to a standardization of what is conceived and planned to be local that is itself homogeneous. The local becomes global in its reproduction of the same 'local' features, while conversely – as is often the case with urban conservation planning, the global may itself be a universalization of what was originally local.

As with many areas of policy, including particularly those relating to the conservation of the natural and built environments, and perhaps also cultural policy more generally, the question as to who is making decisions becomes intertwined with the decisions themselves. In this case, it may be as important to determine the identifier as that which is being identified. Here we return to the issue discussed in various guises in many of the preceding chapters of 'insiders and outsiders'. In Belvedere, national policy is implemented locally and local initiatives and ideas are validated nationally. Potentially, however, this creates the absurd situation in which outsiders define the sense of place of insiders who are informed what their recognizably distinct local identity might be. This circumstance would seem to defeat the initial purpose of the exercise. More subtly, however, as exemplified in a number of the preceding chapters, there is the distinct possibility of an interaction between the two place identities, one projected for external consumption, the other intended for local internal use. Outsiders may seek out aspects of the local identity for various reasons, while insiders similarly adopt the externally projected images of themselves at the local scale. Place-product commodification and branding cannot be separated for a perfectly segmented market. In reality local insiders and non-local outsiders, official policies and local reactions, diverge and converge in a continuous dialectic of the individual and the collective.

To date, about one third of the national land area, and about two-thirds of all urban settlements in The Netherlands have been designated under Belvedere as suitable for long term protection and enhancement for their value to local place identity. This raises the question of the identity of the remaining two-thirds of the country and one third of towns. Have these by default been declared 'identity-poor' or even 'identity-less'? Do the inhabitants feel no sense of place or only that their place has an identity that is less easily recognizable or less valuable than somewhere else? The idea of the existence of an 'identity value surface', which, at least in theory if not in practice, could be mapped, points to some intriguing possibilities with applications in the geography of decision-making. Such questions return the discussion to the differences between the official reliance upon the more universally recognized physical attributes of a region, and the intangibles that contribute most strongly to an individual's unofficial collectively unendorsed sense of place.

It cannot also be assumed that there is one single identifiable collective place identity. If any single idea links all the case studies in this book, it is that society is diverse and these many diversities will result in equally diverse place identities. Like the sense of time transformed into heritage, the user creates place identity. Thus it can be argued that programmes such as Belvedere are quite fundamentally flawed. The identity being sought is a chimera and the process of the search is a serious denial of the social, ethnic and racial diversity of contemporary Dutch society. If there is one single lesson that this book can teach, it is that words such as 'identity', alongside 'heritage', must almost always be pluralized when used in public policy.

There remains a key issue that is rarely posed explicitly in the policy statements such as Belvedere nor indeed in the case-studies presented here, largely because the answer is assumed to be either self-evident or devoid of any useful meaning. Do people need to identify with places? Belvedere and similar policies are driven by the assumption that a distinctive, clear place identity is, if not a necessity, at least beneficial. Specifically, certain benefits, whether economic or psychic, emanate from the officially endorsed 'identity-rich' regions and are conferred upon their inhabitants and other users of such places. Conversely 'identity-poor' areas have no such advantages, which applies both to their economic and cultural commodification. This prompts two conclusions. First, it can be argued that these assumed benefits accrue to collectivities and contribute to collective attributes such as social cohesion or political allegiance, whereas the individual does not automatically receive any benefit. Secondly, it is dangerous to conflate identity with place-identity. Far from being a universal basic human need, it can be argued that much social and even political identification may have no need of place. But, as the case studies in this book have shown, the creation of senses of place from senses of time, and their orchestration through the mediation of official/ unofficial and insider/outsider, remains simultaneously a fundamental psychic and economic necessity. That those twin dichotomies define virtually every incidence of place, identity and time means that the contestation of heritage is also an inescapable human condition.

Index